T0205595

Industry 5.0

Carolina Feliciana Machado · João Paulo Davim
Editors

Industry 5.0

Creative and Innovative Organizations

 Springer

Editors
Carolina Feliciana Machado ⒾⒹ
School of Economics and Management
University of Minho
Braga, Portugal

João Paulo Davim ⒾⒹ
Department of Mechanical Engineering
University of Aveiro
Aveiro, Portugal

ISBN 978-3-031-26234-0 ISBN 978-3-031-26232-6 (eBook)
https://doi.org/10.1007/978-3-031-26232-6

This Springer imprint is published by the registered company Springer Nature Switzerland AG
The registered company address is: Gewerbestrasse 11, 6330 Cham, Switzerland

Preface

Following a period of deep discussion around industry 4.0, the central issue of today is already developing around a new concept, such as industry 5.0. Indeed, if it is true that industry 4.0 is considered to be of great importance from an early age thanks to its ability to increase the levels of efficiency and productivity of the organizations and industries, it is also true that it has a set of concerns, as is the case of the role played by human beings in organizations, as well as the level of unemployment that can be observed as a result of the increasing automation introduced by industry 4.0. In this sense, more recently, we are faced with the so-called 5th industrial revolution, in which the human being assumes a relevant role due to his ability to think, which, by allowing him to more effectively exploit the "intelligence" of software and/or computer applications, allows him to develop solutions with a personalized character. Thus, we are faced with industry 5.0 which, exploiting the potential of research and innovation, helps organizations to become more resilient, sustainable and focused on the human factor.

According to the European Union *"Industry 5.0 is characterized by going beyond producing goods and services for profit. It shifts the focus from the shareholder value to stakeholder value and reinforces the role and the contribution of industry to society. It places the wellbeing of the worker at the centre of the production process and uses new technologies to provide prosperity beyond jobs and growth while respecting the production limits of the planet"*.

In industry 5.0, human beings and machines interact positively to achieve a more sustainable world. This is the challenge that organizations and management face; to the extent that together, human being and machine contribute to the achievement of a wide range of opportunities, making companies increasingly sustainable. More specifically, the challenge that management is facing is then to be able to integrate its employees and technology/machines, thus maximizing the benefits that this interaction allows to obtain. On the contrary to the "traditional" idea that machines substitute the human being, according to industry 5.0, management must be able to highlight the critical role of their collaborators, valuing their intuitive and problem-solving ability of which only the human being is capable, thus making it irreplaceable. If it is true that machines are more robust and more accurate than humans, it is also true that they are characterized by less flexibility and adaptability, as well as critical thinking, characteristics that are only present in humans. Management is therefore required to be able to adapt, adopt the principles of green and lean production, following the digital transition and acting proactively as a way of responding to the challenges that the environment poses to it.

Considering the deep changes and challenges that the organizations' management nowadays face, as a result of industry 5.0, with this book, entitled *Industry 5.0: Creative and Innovative Organizations*, we look to study and understand how todays' organizations and management act in order to more effectively harness the full potential provided by industry 5.0. In other words, in what way today's organizations, as well as their management, lead with the human–robot co-working? In what extent management is able to take decisions related with the organizational issues that are emerging from this interaction? What challenges are posed to the management toward sustainable, human-centric and more resilient organizations? Knowing that industry 5.0 is required to provide a better interaction between humans and machines in order to achieve effective and faster outcomes, to what extent does management develop the necessary measures, policies and practices in order to take advantage of the full potential underlying here?

From the above, it is easy to conclude that today's management is faced with numerous challenges that require it to systematically think strategically in order to more effectively exploit the potential provided by industry 5.0. The growing and continuous technological development has led us, successively, to new management paradigms. In this sense, transforming and adjusting the way in which organizations operate and processes, adapting them to the principles of digitalization; the search for business models that contribute to the use of the least resources to obtain the highest profits; the ability to join man and machine, working together, in order to make the best decisions for the organizations; the ability to develop sustainable policies; and the recognition of the human being value and potential, not only in leverage the potential of technology, but also in exploring its own ideas able to lead to products and/or services that are developed in a personalized way, are only some of the challenges that are gradually being put to management in order to ensure that the organization remains dynamic and competitive.

This book is designed to increase the knowledge and understanding of all those interested in the management and organizations' evolution, resulting from the continuous industrial revolutions that the world is facing, with particular emphasis on the current 5th industrial revolution, better known as industry 5.0, that develop their roles in the different fields of activity like university research (particularly students at the undergraduate level), business, manufacturing, education, health care as well as other service and industrial sectors.

Organized in eight chapters, *Industry 5.0: Creative and Innovative Organizations* looks to cover in Chapter One "University and Education 5.0 for Emerging Trends, Policies and Practices in the Concept of Industry 5.0 and Society 5.0"; while Chapter Two deals with "The Process of Selecting Influencers for Marketing Purposes in an Organisation". Chapter Three discusses the "Personalization of Products and Sustainable Production and Consumption in the Context of Industry 5.0";

Chapter Four covers "Energy in the Era of Industry 5.0—Opportunities and Risks"; and Chapter Five speaks about "Assessing the Drivers Behind Innovative and Creative Companies. The Importance of Knowledge Transfer in the Field of Industry 5.0". Chapter Six deals with "A Brief Glance About Recruitment and Selection in the Digital Age", while Chapter Seven discusses "Conscious Humanity and Profit in Modern Times: A Conundrum". Finally, Chapter Eight focuses "Multigenerational Men and Women and Organisational Trust in Industrial Multinational Firms in Portugal".

Contributing to stimulate the growth and development of each individual in a competitive and global economy, *Industry 5.0: Creative and Innovative Organizations* can be used by academics, researchers, managers, engineers, practitioners and other professionals in related matters with management and business.

The editors acknowledge their gratitude to Springer for this opportunity and for their professional support. Finally, we would like to thank to all chapter authors for their interest and availability to work on this project.

Braga, Portugal   Carolina Feliciana Machado
Aveiro, Portugal   João Paulo Davim

Contents

Editors and Contributors

About the Editors

Carolina Feliciana Machado received her Ph.D. degree in Management Sciences (Organizational and Politics Management area/Human Resources Management) from the University of Minho in 1999, master degree in Management (Strategic Human Resource Management) from Technical University of Lisbon in 1994 and degree in Business Administration from University of Minho in 1989. Teaching in the Human Resources Management subjects since 1989 at University of Minho, she is since 2004 Associate Professor (with Habilitation since May 2022), with experience and research interest areas in the field of Human Resource Management, International Human Resource Management, Human Resource Management in SMEs, Training and Development, Emotional Intelligence, Management Change, Knowledge Management and Management/HRM in the Digital Age/Business Analytics. She is Head of the Human Resources Management Work Group in the School of Economics and Management at University of Minho, Coordinator of Advanced Training Courses at the Interdisciplinary Centre of Social Sciences, Member of the Interdisciplinary Centre of Social Sciences (CICS.NOVA.UMinho), University of Minho, as well as Chief Editor of the *International Journal of Applied Management Sciences and Engineering* (IJAMSE), Guest Editor of journals, books Editor and book Series Editor, as well as Reviewer in different international prestigious journals. In addition, she has also published both as Editor/Co-editor and as Author/Co-author several books, book chapters and articles in journals and conferences.

João Paulo Davim is Full Professor at the University of Aveiro, Portugal. He is also distinguished as Honorary Professor in several universities/colleges/institutes in China, India and Spain. He received his Ph.D. degree in Mechanical Engineering in 1997, M.Sc. degree in Mechanical Engineering (materials and manufacturing processes) in 1991, Mechanical Engineering degree (5 years) in 1986 from the University of Porto (FEUP), the Aggregate title (Full Habilitation) from the University of Coimbra in 2005 and the D.Sc. (Higher Doctorate) from London Metropolitan

University in 2013. He is Senior Chartered Engineer by the Portuguese Institution of Engineers with an M.B.A. and Specialist titles in Engineering and Industrial Management as well as in Metrology. He is also Eur. Ing. by FEANI, Brussels, and Fellow (FIET) of IET London. He has more than 35 years of teaching and research experience in Manufacturing, Materials, Mechanical and Industrial Engineering, with special emphasis in Machining and Tribology. He has also interest in Management, Engineering Education and Higher Education for Sustainability. He has guided large numbers of postdoc, Ph.D. and master's students as well as has coordinated and participated in several financed research projects. He has received several scientific awards and honors. He has worked as Evaluator of projects for ERC—European Research Council, and other international research agencies as well as examiner of Ph.D. thesis for many universities in different countries. He is Editor-in-Chief of several international journals, Guest Editor of journals, books Editor, book Series Editor and Scientific Advisory for many international journals and conferences. Presently, he is Editorial Board Member of 30 international journals and acts as Reviewer for more than 150 prestigious Web of Science journals. In addition, he has also published as Editor (and Co-editor) more than 300 books and as Author (and Co-author) more than 15 books, 100 book chapters and 600 articles in journals and conferences (more than 400 articles in journals indexed in Web of Science core collection/h-index 65+/14500+ citations, SCOPUS/h-index 71+/18500+ citations, Google Scholar/h-index 92+/30500+ citations). He has been listed in World's Top 2% Scientists by Stanford University study.

Contributors

Elias G. Carayannis George Washington University, Washington, DC, USA

Mochammad Fahlevi Bina Nusantara University, Jakarta, Indonesia

Sandra Grabowska Department of Production Engineering, Silesian University of Technology, Gliwice, Poland

Tia Huttula School of Business and Economics, University of Jyväskylä, Jyväskylä, Finland

Heikki Karjaluoto School of Business and Economics, University of Jyväskylä, Jyväskylä, Finland

Eugen Laudacescu Petroleum-Gas University of Ploiești, Ploiești, Romania

Fernando León-Mateos School of Economics and Business, University of Vigo, Vigo, Spain

Lucas López-Manuel School of Economics and Business, University of Vigo, Vigo, Spain

Carolina Feliciana Machado School of Economics and Management, University of Minho, Braga, Portugal;
Interdisciplinary Centre of Social Sciences (CICS.NOVA.UMinho), University of Minho, Braga, Portugal

Ana Martins Graduate School of Business and Leadership, University of KwaZulu-Natal, Westville, South Africa

Isabel Martins School of Management, IT and Governance, University of KwaZulu-Natal, Westville, South Africa

Joanna Morawska Adam Mickiewicz University, Poznań, Poland

Adrian Neacșa Petroleum-Gas University of Ploiești, Ploiești, Romania

Lurdes Pedro Escola Superior de Ciências Empresariais, Instituto Politécnico de Setúbal, Setúbal, Portugal

Marius Gabriel Petrescu Petroleum-Gas University of Ploiești and Romanian Agency for Quality Assurance in Higher Education (ARACIS), Bucharest, Romania

José Rebelo Escola Superior de Ciências Empresariais, Instituto Politécnico de Setúbal, Setúbal, Portugal

Carlos Rodríguez-Garcia School of Economics and Business, University of Vigo, Vigo, Spain

Antonio Sartal School of Economics and Business, University of Vigo, Vigo, Spain

Nara Caroline Santos Silva School of Economics and Management, University of Minho, Braga, Portugal

Sebastian Saniuk Department of Engineering Management and Logistic Systems, University of Zielona Góra, Zielona Góra, Poland

Maria Tănase Petroleum-Gas University of Ploiești, Ploiești, Romania

University and Education 5.0
for Emerging Trends, Policies
and Practices in the Concept of Industry
5.0 and Society 5.0

Elias G. Carayannis and Joanna Morawska

> *We need to change the way we envision both business and society. The old ways have worn themselves out. We are having both a crisis of democracy and a climate crisis. They are both a result of a limited way of thinking.* (Carayannis, 2020, p. 3: http://riconfigure.eu/wp-content/uploads/2020/01/Interview-with-Elias-Carayannis_2020_Final.pdf)

Abstract This chapter focuses on new theoretical constructions that can lead to a more sustainable future, that is a Quadruple/Quintuple Helix approach to innovation and Industry 5.0 and Society 5.0. We see them as a framework integrating and including all the relevant actors of the innovation ecosystems and realms of democratic values in their core. Definitively, there is a need for a new interdisciplinary research between science and engineering with the aim of developing a perfect human-technology collaboration in Industry 5.0. In addition to this, it is necessary to develop and conduct a multi-level analysis of the future university model 5.0. A smart University 5.0 must understand and update the situation inside and outside its boundaries, with a broad perspective of intra-organizational and inter-organizational cooperation. We therefore concentrate on theoretical views and considerations with some practical implications of the aforementioned research concepts and their potential to build a new system of innovation that promotes in a systemic way the open, "glocal", social and digital social innovations for the benefit of people with a key role of science and its social and societal impact. The concept of University 5.0 and Education 5.0 is an attempt to address present ongoing digital transformation and green transitions, and to stimulate the social dimension of universities' missions. In a single university perspective, a micro level would concern the optimization of research and innovation processes. At a meso level, we can assume an analysis of the innovation ecosystem in which the university is located, also including the territorial peculiarities within which cooperative synergies would be

E. G. Carayannis (✉)
George Washington University, Washington, DC, USA
e-mail: caraye@gwu.edu

J. Morawska
Adam Mickiewicz University, Poznań, Poland

developed. Lastly, a macro analysis (completely external agents, such as political, economic, demographic, sociocultural conditions, legal aspects, technology, etc.) can be implemented to support innovation growth based on new routes of a university fully declined? In terms of 5.0 version (Carayannis et al. in J Knowl Econ 13:2272–2301, 2022). This multi-level path is still very relevant to the condition of industry 4.0 towards Industry 5.0/Society 5.0. Starting with a clear vision and mission statement, then translated into strategies and operational plans, can ensure the sustainability of the entire ecosystems of innovation, by taking into account all the Ethical, Legal, and Social Implications (ELSI) involved. Again, given the importance of the social aspects related to the concept of Industry 5.0, Zhang et al. (IEEE Trans Comput Soc Syst 5:829–840, 2018) proposed a paradigm shift from cyber-physical systems (CPS) to cyber-physical-social systems (CPSS). The application of the (eco)logics orbiting around the quintuple helix innovation model (Carayannis and Campbell in Int J Technol Manag 46:201–234, 2009; Carayannis and Campbell in Int J Soc Ecol Sustain Dev 1:45–69, 2010a; Carayannis and Campbell in J Knowl Econ, 2010b) can ensure the continuous interaction of the five dimensions involved: (1) Industry, (2) Government, (3) University (4) Society and (5) Natural Environment, going towards an innovation eco-system design based on a truly human centered "everget-ical" 5.0 paradigm (furthermore, see also Carayannis and Campbell in Innovation systems in smart quintuple helix innovation systems. Springer, Cham, 2019).

Keywords Industry 4.0 · Industry 5.0 · Society 5.0 · University 5.0 · Quintuple innovation helix · Cyber-physical systems · Cyber-physical-social systems · Artificial intelligence

Preface

Between a Little Town in Eastern Poland and a Modern University

The story starts with a Polish carpenter, who gained a basic education in the 20s of the XX century, in a town with a proud tradition of a "royal city" built in 1594, with a first private university in Poland. He took vocational training at the joinery owned by a well-recognized Jew, to start the most prosperous carpentry enterprise in the neighbourhood. The second World War and its aftermath changed everything. In the late 50s, thanks to his good education and a talent for woodworking, he was running a very prosperous business. Times were difficult, but this motivated people to dealing with hardship in a creative way. He constructed! an "automatic" wooden washing machine. It worked for over a dozen years. He was not only an inventor, but also an innovator. With such a creativity, he would probably become a millionaire if it had not been for communist Poland. Unfortunately, the undemocratic system was repressing private initiatives so he had to close his workshop and was forced to work in a public state-owned company until his retirement. He never re-opened his workshop. What does this story tell us about even if it comes from decades ago? That a talented and hardworking man created his washing machine because he had a solid education and because he felt the need for a change. This means that innovation might be natural for people with a strong creativity and motivation if only they got a proper support and incentives. Innovation can also be limited or even hindered if the system does not support the innovators and the unrestrained flow of talent, knowledge and diversity of skills. It also shows that innovation might arise outside, what we call today, a research and innovation system, and the system itself is in constant transformation due to the ongoing social, economic, cultural, political and global changes. And this is where a university should step in as the most appropriate body.

Why is innovation so important if we talk about the future of universities? Because we believe it is a driver of change and if we think about our present times and the challenges we need to face. Innovation can be created without a research leg but through education. Higher education institutions including universities, opened for all generations, might be the place preserving not only academic values but also democratic practices, fundamental rights and freedom of scientific research, inclusiveness, diversity, and values of the whole society. And this is what we should think about today when we discuss the new roles of universities. First of all, we can support innovation through engaging students in new types of start-ups and companies responding to societal needs. We should not strive for patents and other types of closed property, but we should strive for solving problems. To do that we need to prepare the young generation for creativity, problem-solving, team work, critical thinking, challenge based learning, service learning, digital and green skills, etc. We should look for innovations in which nobody claims intellectual property in order to protect it. This opens up ways to social innovations, which responds to public and private values and needs, and is efficient, effective, scalable and targeted. University can respond to societal needs through different types of community engagement like living labs approaches, citizen science, science education, including stakeholders in defining their research and education agenda. A question arises if this will be enough.

This is a very difficult question for many reasons. One of them is that we still believe that a university is something more than only a research and education, and a third or fourth mission. A university represents such values as freedom, autonomy or truth-seeking. This is a community which develops new values and changes the society. Even if we assume now that most of innovation is created by business, we shall not forget that it is created by human beings who probably got their education and developed their creativity at some good university. Science and education create progress understood in many ways. But on the other hand we can still see a distance between science and society. WHAT ??? represents the old model of POWER in which the voice of science is stronger than a voice of society in the sense that society does not actively participate in research and education. The knowledge transmission is usually one step forward. We deliver our values and intellectual outputs to the society not necessarily truly responding to the society needs in the way it is expected. This creates a constant struggle between the expectations and the delivery. But we also want to be clear that this kind of thinking might bring dangerous practices and limit the academic freedom that comes from scientists' choices, research passions and interests. We do not want universities to be pushed to give evidence for their

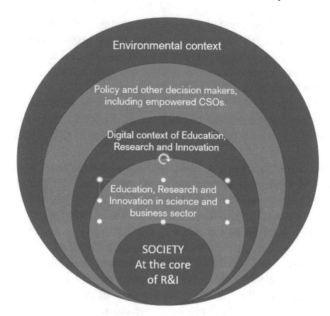

Fig. 1 Society 5.0 in the Quintuple Helix—a reverse perspective. *Source* Authors

existence and to constantly ask for respect and public funding. We need to find a balance between those two POWERS and to establish NEW POWER relations serving the expectations of the society on one hand, and on the other will create new research agendas, questions, boost the creativity and help to build the culture of trust toward science, without hindering the universities' autonomy.

If we are thinking about the new POWER relation on the ground of a regional innovation system, we, as universities, are encouraged to orchestrate this process as institutions that are more reliable then policy makers. However, can we imagine that this process will be inverted in the sense that the society will have the power to structure this process and to be deeply involved in the creation of innovations (Fig. 1)? In other words, the system will be HUMAN-CENTRIC. This sounds difficult because the society is represented by many organisations with different aims, values and needs and the whole system is transdisciplinary. This is also a matter of POWER. There is a fear that if we want to EMPOWER some groups, we need to take POWER away from those who are privileged. Or we want to give POWER to those who may endanger current privileges or whose EMPOWEREMENT will change social relations and transform the whole system.

In this chapter, however, we will try to develop our theoretical views on a new university and education model incorporating the core assumptions of the Society 5.0 and nested into the Quintuple Helix Model of Innovation. In a case study, we will also show evidence that engaging society into the innovation process brings additional values and opens up a new perspective. It also preserves democratic values. When we think about reversing the innovation system and the change in POWER, we can assume that the society should be at the centre of innovation, and not the innovation and economic progress itself. So all the effort towards innovations, all the technology we have, all the POWER that comes from the research and technology should be used to respond to the society needs and values. A great role in this process belongs to universities. A university has potential to deliver a fair distribution of knowledge, power and innovation within society.

Universities used to have great potential to assimilate social needs and collective interests, to play a transformative roles, to open up new strategic channels of collaboration. This potential came from the POWER of authority, tradition, and knowledge. The university "ivory towers" were criticised on one hand, but on the other they created a picture of the temple of knowledge, to which only wise and privileged people had access. This was not necessarily in line with democratic values, but it helped to build authority. For example, the democratisation process, which took place in Poland in the early 90s of the twentieth century, has also opened up access to education and science and created a "wholesale" university. Various consequences of this process we are observing today. We are proud to have a well-educated society, but the quality of education is constantly questioned. In this "wholesale" process we forgot about the real university mission, which is looking for truth and shaping good citizens, and instead we have been selling a product called higher education without thinking about its value chain process and the expiry date. If we are forgetting about those fundamental values, the process of democratisation may become cartoonish. Needless to say, we can experience this on different levels in our modern world. Democracy needs a lot of efforts, wise strategies responding to current challenges, and our ability to distribute power, wealth, prosperity, safety, as well as education in a reflective and responsible way.

To create a more responsible university and innovation system, we need to change and to establish new POWER relations within a university and between science and society. First, we need much more adaptability and flexibility. The world was already in rapid transformation, but the COVID-19 pandemic has accelerated this even more. We need to take advantage of this inflection point so that we do not return to outdated models. Institutions, organisations, companies, universities and all society must transform themselves and embrace the uncertainty and the transformation "in progress". One of the challenges of this transformation is to upskill or reskills continuously. What universities still do is rather RESPONDING and not CREATING and DEFINING FUTURE VISIONS. This lies in hands of big industry. So let's think about the learning which is becoming more digital or hybrid. It affects not only the system of learning, the roles of a teacher and a student, but also the roles of space. Can we create different roles for our buildings, how to organise their space in a way it is adapted to modern students' needs, but are we ready to provide the lifelong learning for our alumni? With an ageing population and economic crisis, companies will hire more and more elderly people and people would probably need to work longer. Do we think about their educational needs? We do not want university to "produce workers" for the labour market, but not replying at all is a serious mistake. It is often said that universities will survive like they have always done. They will, but there is a question at what costs. If we want to preserve our reputation as the anchor of the system, we need first to encourage students to stand in our doors. So we need to open them for our students and the whole society.

What we try to stress in this chapter is that the Artificial Intelligence has a great POWER to transform our societies and it has already changed the POWER relations and influenced modern democracies in many ways. The consequences are still yet to come and our role as universities is to be prepared, protect ourselves from the illegal or not ethical use, to develop and distribute novel technologies in such a way that they will help to build a more resilient and more sustainable future. It is part of the universities' responsibility to transfer those new types of technologies to the society and to use it for the public good. Universities should use their educational mission to EMPOWER people to have better lives, better jobs, and a better health. There are many research projects being developed connecting AI and ICT with the society and the places (local context). This digital context of innovation is becoming more and more important and the challenge is to have a balance between the POWER to change the society and the POWER to breach the democracies. This digital context of innovation might change the way innovation and knowledge are distributed and created within socio-economic systems. Universities should look for new types of innovation which will be both technical and social. That will help to integrate different approaches through new information and technological channels allowing to bind public opinions and voices (e.g. GIS systems in urban development, crowd mapping, crowd sourcing etc.) with social outputs and new types of solutions. The POWER of new ICT tools can lead to a more democratic approaches in managing, transferring and distributing the knowledge from and to the society.

It is often said that we need to focus on societal needs and values. Unquestionably, they are expressed and codified in some way within the Sustainable Development Goals of UN. This globally agreed spectrum of goals ask for urgent actions and solutions developed by different types of stakeholders in co-developed, co-created, co-delivered and co-experimented ways. The Quintuple Helix approach might foster this process as it integrates different perspectives, and sets the stage for sustainability priorities and considerations. In this new context of Society and Industry 5.0, the society is at the core of innovation system. Education, Research and Innovation are delivered and developed by universities and business, which reflects their strong relations in the regional innovation system, and stress the process of life long learning which is being continued at a work place, and stress the need for new paths of flexible learning which should be offered by universities. Those new processes are taking place also in a digital context, and this can help to develop new forms and channels of distribution of E, R and I. Policy and other decisions makers manage and facilitate the system of innovation and since new types of innovation are being developed, including user-driven innovation, open innovation, social innovation, they open up this process and also give new POWER to the society and its representatives, to be engaged in the distribution and creation of new types of innovation. If the needs and values of the society are to be reflected in the E, R and I, they also need to be present on the level of decision making. And finally, Nature is reflected in the environmental context which is needed if new innovation systems are to be delivered to SGD Agenda. Therefore, this asks for new curricula, research, and a new dialogue, which should help to establish new POWER relations on the level of communities, regions, countries and globally.

This last component takes us back to the democracy and POWER relations. The ongoing green and digital transformations lead to novel solutions toward the planet, its biodiversity protection, adaptation to climate change and its mitigation. This in turn may change those relations. The conflicts we observe reflect the conflicts of POWERS and interests of different groups. The environmental justice is something that is calling for new understanding and organisation of democracy. The system has to be rebuilt so that it should include and protect the weakest through de-empowering the strongest, but without large-scale side effects leading to new conflicts. This is what we might call a sustainable transition. It needs much commitment and cooperation. It needs strong leaders and re-building the democratic values all around the globe understood as a right to live (in a healthy environment); justice (environmental justice as well); common good, equality, brotherhood, truth (belief in scientific facts); freedom (from particular interest, from conflicts) etc. This is fundamental for the sustainable development for us and future generations.

We started with a past generation, and now let's look into the future one that will soon be in POWER. They are now in primary or high schools, but soon they will be standing in front of university doors with their dreams and expectations for a good job, a good quality of life and undoubtedly for peace. What we already know about this generation is that they learn and gain knowledge in a different way, using different tools, being stronger in some fields (looking for information), but weaker in others (transforming knowledge into practical skills and solutions), and still needing guidance and mentors. Universities should be ready to empower this new generation and teach them to think in a critical way, to deal with technical disruption, the information' overload, to create trust-based relations with other people, to be able to use the knowledge in both social and work life. We all want them to live in a world of peace and preserved nature, and not to experience the "dinosaurs extinction" again. This is not the POWER of a single person or institution, so universities should intensify the POWER of synergies in e.g., the concept of the European Universities. Those synergies come from different disciplines, sectors, societal groups incorporated into the innovation system of the University 5.0.

1 Introduction

Universities, for their part, in addition to spurring technological progress as before, must additionally be responsible for cultivating literacy among information users through both general curricula and recurrent education, so as to promote the civil society that embodies Society 5.0.[1]

---

[1] Society 5.0, 2018, p. 13.

The social relevance of research & innovation, called responsible research and innovation (RRI), has gained its momentum last decade. In 2014, the Rome Declaration defined RRI as 'the ongoing process of aligning research and innovation to the values, needs and expectations of society. It also stated that 'RRI requires that all stakeholders including civil society should be responsive to each other and take shared responsibility for the processes and outcomes of research and innovation' (Grau et al., 2017). RRI has become a key concept in the international sphere, along with open science, citizen science, sustainable science, science with and for society, participatory research and co-creation. The foundation of this concept arises from the fact that present challenges cannot be simply addressed from a unilateral perspective, but new forms of innovation e.g., social innovation, user-driven innovation or open innovation should be recognised as an important component of this new framework. Those challenges in their dynamic complexity require new cross-scale, cross-domain and action-oriented approaches at the universities. In this chapter we argue that universities need to go beyond their traditional missions and take an active role in a transformative change by working with their communities and creating a real social impact. The question remains what tools and methodologies can be used by universities to maintain the role of the anchor of innovation ecosystems. This is one of the gaps we try to address in this chapter, with focus universities' key missions and their external dynamics, on condition that the new paradigm of knowledge democratisation is built upon the cooperation with non-academic actors. Surely, 'universities are complicated mixtures of different communities with changing power and specific relations with external actors' (Arocena & Sutz, 2021, p. 4). Those new types of relations are reflected in Q2HM and 'only few contributions have explored the connection between the social innovation concept and the QHM framework' (Bellandi et al., 2021, p. 8). In this chapter we address this gap and propose that apart from differences among universities (in terms of their history, relevance, missions, profile, research and education strategies, funding, etc.), their embedment in the regional ecosystem of innovation is one of the key dimensions that can influence their engagement in innovation. We also argue that social innovation, and generally speaking human-centric innovation should be extended to all the missions. We attempt to approach this concern through a theoretical views and a showcase of a TeRRIFICA project.

2   Industry and Society 5.0 and Quintuple Helix

In this chapter we focus on two theoretical constructions which are relevant for understanding the modern process of innovation. The first is the Quadruple/Quintuple Helix framework of innovation ecosystem (Carayannis, 2020, 2021; Carayannis & Morawska-Jancelewicz, 2022; Carayannis & Morawska, 2023; Carayannis & Rakhmatullin, 2014; Carayannis et al., 2021c)??? that which is open, ?? non-liner, co-created, co-constructed and inclusive with civil society organisations and the environment as the active actors. The second is the Society 5.0 and the Super Smart Society

(Breque et al., 2021; Carayannis & Morawska-Jancelewicz, 2022; Carayannis & Morawska, 2023; Carayannis et al., 2020b, 2021b, 2021c; Fukuyama, 2018), which highlights the need to re-think the existing working methods and approaches toward innovation and to focus them on developing human-oriented solutions and social innovation. After Carayannis and Campbell (2009, 2010a, 2010b) and Carayannis et al. (2021a) we might claim that the five dimensions of the Quintuple Innovation Helix (Fig. 2) clearly qualify to relate to themes of Industry 5.0 and Society 5.0, which are manifestations of institutional, cultural, legal, social, political, economic and technological embodiments of the nexus of the government, university, industry, civil society and environmental dimensions.

Industry 5.0 is considered to be a renewed human-centred/human-centric industrial paradigm, starting from the reorganisation of the production processes to the generation of positive implications (both within the business perspectives and all the components belonging to the innovation ecosystem) (Carayannis et al., 2020a, 2020b, 2021a, 2021b, 2021c). Industry 5.0 relies on three core elements: human-centricity, sustainability and resilience, so business true purpose must include social, environmental and societal considerations (Breque et al., 2021, p. 15). The urgency of Industry 5.0 derives by reason of the fact that Industry 4.0 is only at initial stage of the development and the main achievements can be expected not earlier than in mid-twenties. Furthermore, the normative dimensions (responsible/irresponsible,

Fig. 2 The Quintuple Helix (five-helix model) innovation system. *Source* Carayannis and Campbell (2010a, 2010b, p. 62)

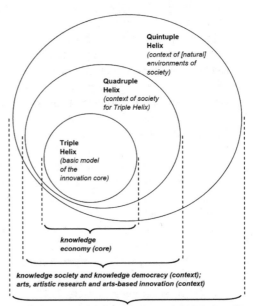

The Quadruple and Quintuple Helix innovation systems in relation to society, economy, democracy, and social ecology.

ethical/unethical) and policies defining global governance of Industry 4.0 are lacking a holistic vision which should take into account the real impact of such issues. As Bartoloni et al. (2021) argue that overcoming some of Industry 4.0 shortages leads to the ever increasing importance of the Society 5.0 paradigm, thus explaining how to design more human–centric solutions, capable of better integrating the I4.0 technologies and human needs. Moreover, after Carayannis et al. (2021b): the "… discussions on Industry 4.0 and Society have, tended to focus on either a dystopian fearful future shaped by the IoT where robots ("CoBots") with AI replace humans, or a future that will invariably be benevolent and prosperous for all with the introduction of the Industry 4.0. Both visions subscribe, however, to technological determinism (evolution in organizational behavior, acceptance of robots in the workplace, evolution in organizational structures and workflows, evolution in work ethics, discrimination against robots or people, privacy and trust in a human–robot collaborative work environment, education and training, redesign of workplaces for robots) (Carayannis & Campbell, 2021), and as the emergence of Industry 4.0 and its societal shaping and impacts are preordained and inevitable, they do not yet acknowledge the need to broaden the understanding of Industry 4.0 outcomes and its multiple possible futures in society.

Society 5.0 (Super Smart Society) is a new guiding principle for innovation developed in Japan last decade, based on the convergence between cyberspace and physical space, and enabling to use the Artificial Intelligence to perform or support the work and adjustments which humans have done up to now (Fukuyama, 2018). Society 5.0 focuses on human beings with the aim to involve a wide variety of actors who in the past only participated in non-visible ways (e.g. women and young people) and it creates a space for accommodating various bottom-up ideas. Society 5.0 calls for "systemization" of services and projects, more advanced systems, and coordination between multiple systems—thus aiming to serve as a Smart Bridge between the techno-centric and human-centric perspectives (Carayannis & Morawska, 2023). Society 5.0 considers social capital as its key asset. It is not only concerned with environmental issues but it also uses the threefold analytical paradigm (structural transformation, technological innovation, and quality of life) to explore how to minimise a whole range of social costs and how to boost productivity (Matsuoka & Hirai, 2018, p. 34). The Super Smart Society is built upon delivering the targeted and personalised, just on-/in-time solution to the people with the aim to provide healthy and safe environment and to promote people's well-being. It is still a vision, directive or goal and not reality. Yet, it opens a new perspective to understanding and utilising the technological advancement and digital transformation for the benefit of society (Carayannis & Morawska-Jancelewicz, 2022; Carayannis & Morawska, 2023; Rego et al., 2021). The vision of Society 5.0 requires that we should think about two kinds of relationships: the relationship between technology and society and the technology-mediated relationship between individuals and society (Society 5.0 A People-centric Super-smart Society, 2018, p. 5). Society 5.0 and Industry 5.0 represent the convergence with Quadruple and Quintuple Innovation Helix frameworks as they emphasise that, over the medium to long term, true and transparent democracy constitutes a sine qua non for smart, sustainable and inclusive growth (Carayannis et al., 2020b, 2021b, 2021c; Carayannis & Campbell, 2021). By referring to the concept of "Society 5.0", Carayannis et al. (2020b, pp. 3–4) explain furthermore: "At the basis of this broadening, the idea of *Society 5.0* (or "Super Smart Society") is defined. This prototypical philosophy originated in Japan and was presented as a core concept in the "Fifth Science and Technology Basic Plan" by the Japanese "Council for Science, Technology and Innovation", and approved by Cabinet decision in January 2016 (Serpa & Ferreira, 2019). It was identified as an overall growth strategy for Japan, and was reiterated in "The Investment for the Future Strategy 2017: Reform for Achieving Society 5.0". In essence, Society 5.0 tries to provide a common societal infrastructure for prosperity based on an advanced service platform. Industry 4.0 follows society 5.0 to a certain extent, but while Industry 4.0 focuses on production, Society 5.0 aims to put human beings at the center of innovation, taking advantage of the impact of technology and the results of industry 4.0 with the deepening of technological integration in improving quality of life, social responsibility and sustainability (Carayannis et al., 2022). This innovative perspective is not restricted to Japan, as it has points in common with those of the UNDP SDGs ("United Nations Development Program" "Sustainable Development Goals" (www.undp.org). Furthermore, unlike the concept of Industry 4.0, Society 5.0 is not constrained only to the manufacturing industry, but it solves social problems with the help of integration of physical and virtual spaces. In fact, Society 5.0 is the society where the advanced IT technologies already discussed (IoT, robots, artificial intelligence, augmented reality, etc.) are actively used in people's common life, in industry, health care and other spheres of activity not for the progress, but for the benefit and convenience of each person (Carayannis et al., 2021b, 2022; Fukuyama, 2018).

In designing this transformation, universities can function as core bases of value creation, and become places where transformation is prototyped with the cooperation of multiple stakeholders (Hamaguchi, 2020, p. 104). Society and Industry 5.0 both reflect fundamental shifts of societies and economies toward a new paradigm to balance economic development with the resolution of social and environmental problems, and to tackle challenges associated with human–machine interactions and skills matching (Breque et al., 2021; Carayannis & Morawska, 2023).The goal of this paradigm is to concentrate on new value creation in society and economy through innovations focused on the provision of products and services adopted for diverse individual needs. In this framework, "Society 5.0 recognizes innovations, especially social innovations, and innovativeness of all stakeholders in society as necessary preconditions for the development of information society into human-centered society, based on socially responsible society composed of individuals and their organizations" (Potočan et al., 2021, p. 808).

3   University 5.0

"Recently announced the EU Digital Strategy wants to ensure not only that Europe is a global digital player, but also that the EU leads in making sure that technology works for all, and that we live in an open, democratic and sustainable digital society"

(Correia & Reyes, 2020, pp. 4–28). Universities need to find a *sustainable equilibrium between ecological, economic and social concerns, navigating the digital transition and dealing with (geo)political uncertainty* (Jørgensen & Claeys-Kulik, 2021, p. 10; EUA, 2021). Our theoretical considerations lead to the model of socially and digitally engaged universities which embrace new university roles in the ecosystem of innovation, understood as a multilayer framework in which institutions interconnect to develop and share information and knowledge required for the development of new innovation processes (Carayannis & Morawska-Jancelewicz, 2022; Carayannis & Morawska, 2023). In this new ecosystem led by universities, innovation emerges as a result of the collaboration and co-creation among all actors of innovation. This approach emphasises also the position and roles of local and public actors, and the public policy challenge is to provide the means and instruments to transform traditional environments in an innovative ecosystem of innovation (Costa & Matias, 2020, p. 2).

In our model (Fig. 3) universities are envisioned as prototyping places for social and digital transformations (SDT) and creating POWER CAPITAL. We do not focus on new technologies themselves, but rather on policies and visions related to the new roles of universities in Industry and Society 5.0 within Q2HM. This model has two dimensions. The first refers to a strong academic leadership that recognizes the value of diverse networks which extend beyond their zones of proximity, familiarity and competence based on a dialogue and influence. It also reflects the power of scientists and students to become change agents (Carayannis & Morawska, 2023). We agree with Blewitt (2010, p. 396) who claims that "with information growing by the second, knowledge expanding exponentially and wisdom still in short supply, applying new digital technologies to the sustainability imperative, requires a transdisciplinary synthesising mind and a higher educational specialist who helps students become generalists (Rego et al., 2021). The second dimension refers to the engaged and inclusive society, playing an active role in the innovation ecosystem. We might call it a Super Smart Society in Society 5.0, where value is generated not from clusters of tangible assets, but rather from knowledge spaces where data and information are gathered, and then deciphered and deployed through knowledge (Deguchi et al., 2018, p. 11).

In our vision University 5.0 needs to:

- create proper structures and mechanisms supporting the development and implementation of social/digital innovation (such mechanisms could be financial incentives, but also acknowledging the innovative initiatives of scientists and students; promoting innovative culture, creation of innovative co-working space like e.g. Fab Labs);
- extend (digital) social innovation (DSI) to all the missions (e.g. promote DSI within curricula, short courses, support and develop DSI start up or spin offs, include DSI into research agenda—as theoretical but also practical concepts, support research dedicated to solving grand global and local challenges related to SDGs);

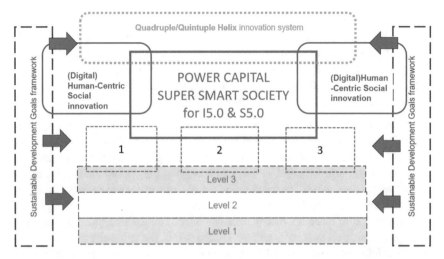

Fig. 3 Socially and digitally engaged model of university in Society 5.0. Level 1: societal and sustainable priorities incorporated into university strategy and missions. Level 2: university leadership focused on protecting academic freedom and autonomy and on future oriented challenges related to digital, green and social transformation. Level 3: green and digital culture and literacy of students, academics and administration promoted and integrated with optimal digital tools and infrastructure. 1. Key mission: education (intelligent, flexible, inclusive, accessible and adaptive learning systems for all generations). 2. Key mission: research (trans and inter-disciplinary). 3. Key mission: Third/Fourth missions or public engagement (cross sectoral and multi-actor). 4. Building POWER CAPITAL/SUPER SMART SOCIETY for Industry 5.0 and Society 5.0 through co-creation with stakeholders and communities for sustainability and Digital (Social) Innovation. *Source* Carayannis and Morawska-Jancelewicz (2022) and Carayannis and Morawska (2023)

- incorporate the societal and sustainability priorities in a systematic way and by this to play an active and leading role in Q2HM (include SDGS into university strategy and mission; monitor, evaluate and validate sustainable practises, link research groups but also administrative structure around those priorities with the aim to overcome silos and competing policies);
- embrace trans- and interdisciplinarity in research and education (e.g. link research groups around digital and green twin, promote students initiative embracing skills from different disciplines);
- promote cross-sector and multi-actor collaboration (e.g. strengthen the involvement of business and civil society organisations in education and research mobility of staff and students through e.g. Living Labs, including virtual, recognise other than publications outputs and measures);
- incentivise utilisation of AI wherever it can offer benefits to the economy and society (e.g. promote intelligent learning like games, geographic information systems and create new flexible, inclusive, accessible and adaptive learning systems for all generations; deliver tailor-made solutions through social/digital innovation);

- promote new curricula focused on green, digital, quantitative and ethical skills necessary to ensure the effective and appropriate utilisation of AI;
- digital transformation and AI curricula embed in Responsible Research and Innovation approach with the aim to anticipate negative impact of AI;
- focus its research, education and innovation more on social well-being and the quality of life (Carayannis & Morawska, 2023).

The model relies on three fundamental pillars. The first assumes that both societal and sustainable priorities should be incorporated into university strategy and missions. The second, requires a strong leadership protecting the core academic values, but also future oriented values, understanding the present challenges as part of building power capital. The third, embraces the new green and digital culture and literacy of students, which impacts both research, teaching and public engagement. They are linked to basic university missions which incorporate those fundamental assumptions and new university culture. Owing to this, the university promotes intelligent, flexible, inclusive, accessible and adaptive learning systems for all generations leading to a new power capital and trans and inter-disciplinary research as well as multi-actor and cross-sectoral public engagement (Carayannis & Morawska, 2023). They are all interrelated and through (digital) human-centric innovation they create a new innovation ecosystem which is sustainability-oriented and embedded in the Q2HM frameworks. This approach allows universities to contribute more strategically, directly and effectively to present global and local challenges around all the university missions (Carayannis & Morawska-Jancelewicz, 2022). As Carayannis (2020) suggests, universities should move from tactical fragmentation to strategic integration and promote a new mindset related to developing new solutions, which he calls six rules of thumb. One should ask if they are (1) Ethical, (2) Efficient, (3) Effective and if the they are (4) Environmentally sustainable, (5) Socially sustainable and (6) Financially sustainable. These six rules of thumb should be linked with four ways to evaluate projects, policies and solutions: metrics, measuring, management and monitoring. We assume that the function and role of university and its staff in the (digital) social innovation process is based on three pillars: (1) A university provides knowledge (existing or developed as part of the cooperation with the environment) which supports the creation of innovation. (2) A university shares its tangible and intangible assets. (3) A university supports (digital) social innovation development by advising social innovators and involving interested parties. Knowledge and support resources may be provided at various stages of creating social innovation and in different dimensions (Benneworth & Cuhna, 2015, pp. 10–12). The fourth mission concept is particularly relevant here as it puts emphasis on a university roles in sustainable development and it is defined by Riviezzo et al. (2019, p. 31) as 'the promotion of social, cultural and economic development of the host community, which, in a very broad sense, leads to the argument that university should contribute also to the quality of life as perceived by the community itself. We believe that integrating both green and digital transition in the university missions leads to the development of (digital) social innovation and to the more open and human-centric innovation ecosystem (Carayannis & Morawska-Jancelewicz, 2022; Rego et al., 2021).

Digitally-enabled collaboration with the actors of innovation ecosystem can also catalyse research and innovation that addresses societal challenges and increases European competitiveness and is one of the pillars of European Commission Open Science initiative, in which research is collaborative, open, responsive, participatory (Carayannis & Morawska, 2023). "It aims to share knowledge and tools as early as possible between researchers in different disciplines and with society at large. It includes, but goes well beyond the concept of open access and open data. In addition to making research cultures more open, it actively seeks to invite and engage stakeholders and citizens from beyond the academic realm into research and innovation processes, for example through public engagement and citizen science (Owen, 2021, p. 5). Those new practices may lead to creating "the ecosystem that will consist of a dynamic, interactive network embedded in an innovation mindset, an interactive set-up focused on knowledge creation and diffusion. These ecosystems might be virtual due to the digital transformation we are facing globally; however, they need some grounded hub as members need to physically meet to interact and co-create, to develop new ideas benefiting from their multidisciplinary skills and competences" (Costa & Matias, 2020, p. 3). To summarise, universities need to *move beyond the future*. They have the tools, they just need to use them. In the next section we are giving an example of this new approach implemented within a EU Horizon2020 project that is based on the co-creation and Living Lab approach. It is also rooted in the Quintuple Helix model and, in a way, through the use of the crowd-mapping tool linked to the vision of the Society 5.0.

4   From Theory into Practice of University 5.0

TeRRIFICA, Territorial RRI fostering innovative climate action.[2]

As stated on a TeRRIFICA project website (https://terrifica.eu/) "Climate change is the defining challenge of our time. Mitigating its impacts and adapting to changes already taking place or impossible to be avoided will require fundamental changes to societies and behaviours all over the world—as well as scientific breakthroughs, both technological and social. The EU intends to remain at the forefront of the fight against climate change and the necessary transition to low-carbon, sustainable and climate-resilient societies". With this objective, in 2018 the European Commission published an evaluation of the EU strategy on adaptation to climate change and also its strategic vision for a new EU long-term strategy for reducing greenhouse gas emissions, setting out clear priorities to achieve a net-zero carbon economy in 2050.

---

[2] https://terrifica.eu/about-terrifica.

"EU-funded research, science and innovation have underpinned both reports, and will keep playing a crucial role in our efforts to tackle climate change and here the EU will continue to lead. We have put the climate at the heart of Horizon Europe—the EU's next research and innovation programme: If we want to achieve a net-zero carbon economy by 2050, more and better focused R&I is a necessary condition to reach this target and to maintain our standard of living"—Climate Change Adaptation. Directorate-General for Research and Innovation, European Commission.

It is in this context that the TeRRIFICA project emerges. Starting on January 2019 and with duration of three and a half years, the TeRRIFICA project ????? set up tailored roadmaps and key performance indicators for the implementation of the developed methodologies and climate change adaptation and mitigation activities in regional practice. The project ???? has been implemented by the following partners: Wissenschaftsladen Bonn (Bonn Science Shop), Germany—project leader; Association of Catalan Public Universities, ACUP, Barcelona, Spain; Sciences Citoyennes, Paris, France; Adam Mickiewicz University, Poznań, Poland; University of Vechta, Germany; Association Education for Sustainable Development, Minsk Belarus and Center for the Promotion of Science, Belgrade, Serbia. A customised capacity building for the different stakeholder groups has been offered. Through workshops, regional and international summer schools, TeRRIFICA aimed to empower local people, with a particular focus on regional authorities and policy makers, and have developed adequate solutions together with them. Field trips to local and regional promising activities related to research and regional innovation, and broader stakeholder engagement with feedback loops ?? have been organised. Through co-creative multi-stakeholder approaches, participants had the opportunity to expand their knowledge around climate change and innovative climate actions, and to identify opportunities, drivers and barriers of implementation. Activities took into account challenges for the acceptance and feasibility, technological and regulatory constraints in six pilot regions. One of them was Poznan Agglomeration, located in the western part of Poland in the center of Wielkopolska voivodeship (analogous to a province). It comprises Poznan and the 17 neighbouring communes. The agglomeration covers an area of 2162 km^2 and has over 1 million inhabitants. It is one of the most important economic and academic centers in the country, characterized by a buoyant and developed labour market, diversified economic structure, established transportation network, and a high level of attractiveness for tourism. Thanks to the diversified structure of the social and environmental system, the Poznan Agglomeration is an interesting area for the analysis of the functioning of the stakeholder network in the context of cooperation with climate change adaptation and mitigation (Fagiewicz et al., 2021). Throughout three years of project implementation, the team from the Faculty of Human Geography and Planning at Adam Mickiewicz University, Poznan, has led the process of developing Climate Change Adaptation Plan for this region. The methods used included e.g. desk research, Delphi study interviews, social methodology lab workshops, national summer school addressed to the young generation, and service learning with students. The co-creation team consisted of representatives of all the four helices of Q2HM. One of the innovative aspects of the project was to include citizens' opinions and voices through the crowd-mapping tool developed by the university researchers in a co-creation process with the relevant actors. Crowd-mapping is a subtype of crowdsourcing, by which aggregation of crowd-generated inputs such as captured communications and social media feeds are combined with geographic data to create a digital map to bring local issues to the attention of public services. The information can typically be sent to the map initiator by SMS or by filling out a form online, and are then gathered on a map online automatically. Crowdmaps are an efficient way to visually demonstrate the geographical spread of a phenomenon (Churski & Kaczmarek, 2022). The tool is dedicated to the identification of green, grey and blue infrastructure linked to climate change and concrete space. It supports the indication of the places on the map where users have observed positive and negative phenomena linked to climate change and environment. This also helps also to identify the local/ regional key players and stakeholders involved in climate action. The tool is based on the "learning by doing" approach, as through the process of the mapping the users also encouraged to learn. As a result of the crowd-mapping the so called "hot spots" were identified in the agglomeration, that is places that require intervention or the places that had already positively replied to the adaptation. The process of preparing the Adaptation Plan was also supported by the organisation of over twenty workshops for citizens from all the communes. The Plan includes an analysis of hazards resulting from climate phenomena and the results of crowd mapping, taking into account spatial differences related to the features of the environment and forms of development. Against this background, resources and activities for adaptation and mitigation to climate change were identified. The project succeed to achieve a systemic approach to organizing the adaptation and mitigation process in the region. It also managed to increase the openness of science and research in this area to social needs, joint development of research agendas, promotion of civic science, i.e. citizen science according to the Responsible Research and Innovation concept.

The team in Poznan worked through the Living Lab method, which is an example of the growing bottom-up movement at modern universities. It is a response to ongoing societal transformations directed towards a more sustainable future. It requires a transdisciplinary approach, integrating researchers and users, a critical and self-reflexive research approach which relates societal with scientific problems, produces new knowledge by integrating different scientific and extra scientific sights and contributing to both societal and scientific progress. Living Labs is a space where university community of staff and students comprises various roles of researchers and problems being researched, as well as the educators and those being educated (Verhoef et al., 2020, pp. 138–139). As a consequence of project activities, the Living Lab Education for Climate was established with the aim to integrate local schools, communities and other actors in the process of practical implementation of the goals developed within the Adaptation Plan for Poznan Agglomeration. HETEROGENEITY.

The other important aspect of TeRRIFICA was also to involve students in the process. Naturally, they were one of the most active groups identified in a crowd-mapping. The second activity addressed to students was a national summer school "Map the Climate" where students worked on the solution addressed to concrete "hot spots" with the support of the mentors coming from local communities. This gave a chance to exchange different opinions and needs and to confront their skills with real-life problems. Moreover, within their obligatory classes in the spring semester at the faculty, over twenty students were developing the practical solutions aiming at adaptation and mitigation to climate change of the campus space. They had a chance not only to use their practical knowledge in the real environment but also to acquire soft skills like team work, creativity, or critical thinking. The important aspect of those classes was that the students knew the space very well, but still when looking at it from a perspective of climate change challenges, were encouraged to think "out of the box" and to feel empowered to change their closest surrounding. As a result, the project called "The Climate Garden" has been approved by the university authorities for implementation in the following years. The main idea of the garden is to create a kind of climate shelter at the campus, connecting nature based solutions with the integration area for students.

The final step of the students' involvement in TeRRIFICA was the international summer school in Barcelona in September 2022, where a group of over forty students coming from various European countries worked together. During the first day, the participants had a chance to get to know each other and discuss with experts about climate change adaptation and mitigation measures, and about the development of citizen science projects. They also got to learn from different case-studies from the Climate-ADAPT project and prepared presentations in groups. During one of the introductory keynotes, the TeRRIFICA partners presented the activities and the work they had conducted in the different Pilot Regions. The day ended with a walking tour with Elena Lacort, from the Climate Emergency and Environmental Education Service of the Barcelona Metropolitan Area. The group visited two designated climate shelters from the city of Barcelona, Ateneu el Poblet and the Sagrada Familia gardens, and discovered the actions that the Barcelona Metropolitan Area is implementing for climate adaptation and mitigation. During the 2nd day students were divided into groups according to their personal interest: gender studies, urban planning, circular economy and citizen participation. For this activity, we had four experts, including Dr. Hyerim Yoon and Dr. Sergi Nuss, from University of Girona, Dr. Tomasz Herodowicz, from Adam Mickiewicz University (Poznan), and Norbert Steinhaus, the TeRRIFICA project coordinator. After a short presentation, the participants went on tours with the experts to explore how those topics are actually represented in an urban space such as Barcelona. Having that valuable information in mind, they went on to put together the main ideas of each topic and to get started with the "Future workshop". The ideas consisted of different phrases: critical analysis phase, visionary phase and implementation phrase. As a part of the activity, the students had to develop an action plan and to present it in a creative format of their choice. As a result, each group presented their climate visions through performances, TV shows, and videos, which helped all the participants and experts collaborating in the Summer School imagine the paths people can follow to achieve a brighter future. The impact of the International Summer School was presented during the Final Conference: TeRRIFICA & RRI2SCALE in Belgrade on the 23rd of November 2022 (see https://terrifica.eu/about-terrifica).

5   Discussion and Conclusions

The model and the approach to a modern University 5.0 presented in this chapter is rooted in the "Mode 3" type university or higher education institution which would represent (and does represent) a type of organization or system which seeks creative ways to combine and integrate different principles of knowledge production and knowledge application (exemplified by Mode 1 and 2), while, at the same time, encouraging diversity and heterogeneity (Carayannis & Campbell, 2009, 2010b; Carayannis et al., 2020a, 2021b, 2021c). Emphasizing again a more systemic perspective for the Mode 3 knowledge production, a focused conceptual definition may be as follows (Prainsack et al., 2012, p. 49): Mode 3 "… allows and emphasizes the co-existence and co-evolution of different knowledge and innovation paradigms. In fact, a key hypothesis is: *The competitiveness and superiority of a knowledge system or the degree of advanced development of a knowledge system are highly determined by their adaptive capacity to combine and integrate different knowledge and innovation modes via co-evolution, co-specialization and co-opetition knowledge stock and flow dynamics"* (Carayannis & Campbell, 2019; Carayannis et al., 2022). Analogies are being drawn and a co-evolution is being suggested between diversity and heterogeneity in an advanced knowledge society and knowledge economy, political pluralism in democracy (knowledge democracy), and the quality of democracy or knowledge democracy. The "Democracy of Knowledge" refers explicitly to this overlapping relationship. As is being asserted: "The *Democracy of Knowledge*, as a concept and metaphor, highlights and underscores parallel processes between political pluralism in advanced democracy, or knowledge and innovation heterogeneity and diversity in advanced economy and society. Here, we may observe a hybrid overlapping the knowledge economy, knowledge society and knowledge democracy" (Carayannis & Campbell, 2010b, 2014, 2021; Carayannis et al., 2022). DOBRZE HETEROGENEITY.

Universities or higher education institutions of a "Mode 3" type of system are designed to enable a "basic research in the context of application". This aligns with qualities of non-linear innovation. Governance decisions in or on higher education should be based on understanding and sensitivity to the particular Mode in which the organization operates, either Mode 1, Mode 2 or Mode 3, and where the universities as drivers of knowledge and the anchors of innovation play a crucial role in orchestrating the process of innovation, and are pursuing the change (Goddard et al., 2016). The concept of "epistemic governance" emphasizes THE FACT that the knowledge conceptions underlying knowledge production and knowledge application (innovation) are addressed with strategies, policies and measures that ensure quality and continuous quality improvement (Carayannis et al., 2022). Epistemic governance is referring explicitly to the "underlying understandings" that are underlying the structures and processes of an organization. Related to this is the proposed Fractal Education, Innovation and Entrepreneurship (FREIE) organizational governance design (Carayannis & Campbell, 2010b), moving from tactical fragmentation into strategic integration in Europe and beyond. There are a few factors important within this approach: the context matters (institutional, socio-economic, regulatory/legal, socio-technical and cultural); horizon matters (so we propose a longer-term horizon and not a short-term policy making (more than ten years at least); policy matters: on a regional/continental level and piloting new solutions by chosen universities, which is exactly in line with a concept of the European Initiatives. We would architect a university as a flexible open-learning, open knowledge and open-innovation ecosystem which would consist of a network of mutually completing and reinforcing, Trans-disciplinary Research, Education and Innovation Centres (TREICs). Each Centre would have as its charter DNA the Quadruple/Quintuple Innovation helix philosophy, and thus would be organically intertwined with other government, university, industry and civil society entities locally, regionally and even globally. Those centres would be the nodes of the FREIE systems whose "blood" would be knowledge and its circulation, taking place via the pro-active and strategic targeted initiatives, as well as intentionally triggering *"happy accidents"* of strategic knowledge serendipity and arbitrage value. Each centre and a university per se would be considered and leveraged as *"innovation diplomacy ambassadors"* and this means a cross-cutting and cross-leveraging set of visions, missions, strategies and tactics involving relevant ministries and a EU Commission (Lecture at EU-US Science and Technology Council, Vienna, Austria, 2011—http://archive.euusscciencetechnology.eu/uploads/docs/CARAYANNIS_BILAT_2011_final%20(2).pdf).

In a recent interview, Carayannis (2020, 2021) coined the following metaphor: *"**Democracy and the Environment are Endangered Species**"*. In a certain way, the contemporary world may be seen as an unfolding race or as a competition of "Developed Democracies versus Emerging Autocracies" (Carayannis & Campbell, 2014). The concept and theory of the Quadruple and Quintuple Helix innovation systems is based on democracy and ecological sensitivity. "Democracy as an Innovation Enabler" emphasizes a co-evolution of democracy (knowledge democracy) with knowledge and innovation (Carayannis & Campbell, 2021). The approach of Quadruple and Quintuple Helix innovation systems provokes with the following two propositions:

1. Without a democracy or knowledge democracy, the further advancement of knowledge and innovation are seriously constrained. In this sense, knowledge and innovation evolution depend on democracy and knowledge democracy.
2. Ecology and environmental protection represent a necessity and challenge for humanity, but they also act as drivers for further knowledge and innovation (this should lead to a win–win situation for ecology and innovation) (Carayannis et al., 2022).

We stress that the transformation of Industry 4.0 will destroy labor, and the transformation of Industry 4.0 will create new labor, so finally there even may be more (new) labor. This requires, however, to reorganize labor and education in innovative and progressive approaches, so that the net gain of new labor has the full potential to even outpaces the losses of old labor. The competence of persons, people and humans must be developed further and further, to prevent labor from being replaced by automation effects or by artificial intelligence, liberating creativity, enabling inventiveness and driving innovation and entrepreneurship (see for instance: https://www.amazon.com/Leading-Managing-Creators-Inventors-Innovators/dp/1567204856).

Crucial are here multi-facetted competences, where disciplinary professional knowledge is being augmented and recombined with interdisciplinary and transdisciplinary skills and competence (for this also the metaphors of "T-competence?" and "M-competence?" are being used). Creativity and creativity skills are crucial in driving innovation, which again is advancing the evolution of knowledge society, knowledge economy and knowledge democracy. Arts and artistic research represent crucial components in an advanced innovation system (Carayannis et al., 2022).

Artificial intelligence will not replace human intelligence, but artificial intelligence will complement human intelligence. However, the challenge is to organize labor (and the economy, society and democracy) in a way, so that human intelligence is using artificial intelligence for the purpose of supporting (and carrying higher) human intelligence and human labor. Therefore, the idea is to speak more of a co-evolution of artificial intelligence and of human intelligence, but where the humans are in the position of control and sovereign decision-making (also expressed in the metaphor of a "Centaur Intelligence"). Artificial intelligence can provide assumptions and guidance, however, the humans are the ones who are making the decisions or who engage in "(making the) decision-making". There is this understanding that advanced knowledge manifests itself in a diversity of knowledge modes and innovation modes, and this pluralism of knowledge also requires a political pluralism, which is a clear characteristics and component of democracy.

Democracy as an innovation enabler, or the quality of democracy as an innovation enabler, emphasize the connectedness and interconnectedness of (a) knowledge development and of (b) democracy development and democracy evolution. In reference to the example and metaphor of a society of free women and free men in ancient Greece (the democratic polis in Athens), we can speculate, how in Industry 4.0 the artificial intelligence and other advanced technological means could be used and can be used and utilized to carry out the (boring) standard work, whereas persons, people and humans then are focusing more on the interesting work. This we may phrase and paraphrase as a type of Renaissance of (interesting) labor in the Age of Knowledge and Innovation. So what are then the new (and old) forms of entrepreneurship and of creative innovation in Industry 4.0 (or Industry 5.0 in a later phase), what can artificial-intelligence-based entrepreneurship possibly mean? What Industry 4.0 really needs and requires is a ?? Democracy 5.0. If there is Art and Democracy, we should also think about the Art of Democracy (Carayannis et al., 2022).

In the future era of Society 5.0, cybernetics will meet with "Evergetics", as the emerging postnonclassical science of intersubjective management processes in the society. Evergetics in Greek (Ευεργέτης) means "benefactor" and already in its etymological origin we recognize an orientation for "good actions" in management processes and decision-making. In fact, the author of this neologism defined it as «... the science of management processes organization in a developing society, each member of which is interested in augmenting his/her cultural heritage he/she is producing, which entails a raise of cultural potential of the society as a whole and, as a consequence, an increase in the proportion of moral and ethical managerial decisions and corresponding to them benevolent actions in public life. It is clear that to ensure that implementation of Society 5.0 is not just a political-ideological concept, it is necessary to integrate several dimensions, such as innovation policy (from government side), entrepreneurial spirit (from society side), entrepreneurial skills (from civil society and institutions) and so on (Carayannis et al., 2020b, 2021b).

Definitively, there is a need for new interdisciplinary research between science and engineering with the aim of developing a perfect human-technology collaboration in Industry 5.0. In addition to this, it is necessary to develop and conduct a multi-level analysis, which takes into account three levels of framework: *macro*, *meso* and *micro*. A smart University 5.0 must understand and update the situation inside and outside its boundaries, with a broad perspective of intraorganizational and interorganizational cooperation. In a single university perspective, a micro level would concern the optimization of research and innovation processes. At a meso level, we can assume an analysis of the innovation ecosystem in which the university is located, also including the territorial peculiarities within which to develop cooperative synergies. Lastly, a macro analysis (completely external agents, such as political, economic, demographic, sociocultural conditions, legal aspects, technology, etc.) can be implemented to support innovation growth based on the new routes of a university fully declined in terms of 5.0 version (Carayannis et al., 2022). This multi-level path is still very relevant to the condition of industry 4.0 towards Industry 5.0/Society 5.0. Starting with a clear vision and mission statement, then translated into strategies and operational plans, it can ensure the sustainability of the whole ecosystems of innovation, by taking into account all the Ethical, Legal, and Social Implications (ELSI) involved. Again, given the importance of the social aspects related to the concept of Industry 5.0, Zhang et al. (2018) proposed a paradigm shift from cyber-physical systems (CPS) to cyber-physical-social systems (CPSS). The application of the (eco)logics that orbit around the quintuple helix innovation model (Carayannis & Campbell, 2009, 2010a, 2010b) can ensure the continuous interaction of the five dimensions involved: (1) Industry, (2) Government, (3) University (4) Society and (5) Natural Environment, going towards an innovation eco-system design centered on a truly human centered, "evergetical", 5.0 paradigm (furthermore, see also Carayannis & Campbell, 2019).

References

Arocena, R., & Sutz, J. (2021). Universities and social innovation for global sustainable development as seen from the south. *Technological Forecasting & Social Change, 162*, 120399.

Bartoloni, S., Caló, E., Marinelli, L., Pascucci, F., Dezi, L., Carayannis, E., Revel, G. M., & Gregori, G. L. (2021). Towards designing Society 5.0 solutions: The new Quintuple Helix— Design thinking approach to technology. *Technovation*. https://doi.org/10.1016/j.technovation. 2021.102413

Benneworth, P., & Cunha, J. (2015). Universities' contributions to social innovation: Reflections in theory & practice. *European Journal of Innovation Management, 18*(4), 508–527.

Bellandi, M., Donati, L., & Cataneo, A. (2021). Social innovation governance and the role of universities: Cases of quadruple helix partnerships in Italy. *Technological Forecasting & Social Change, 164*, 120518.

Blewitt, J. (2010). The green campus is also a virtual one. *International Journal Environment and Sustainable Development, 9*(4), 392–400.

Breque, M., De Nul, L., & Petridis, A. (2021). *Industry 5.0, towards a sustainable, human-centric and resilient European industry*. European Commission, Directorate-General for Research and Innovation.

Carayannis, E. G. (2020). Democracy and the environment are endangered species, interview with Dr. Prof. Elias Carayannis by Charlotte Koldbye. RiConfigure. Available from https://www.riconfigure.eu/wp-content/uploads/2020/01/Interview-with-Elias-Caraya nnis_2020_Final.pdf. Accessed September 15, 2021

Carayannis, E. G. (2021). Sustainable development goal 8—An interview with Elias Carayannis. Available from https://www.springer.com/journal/13132/updates/19771722. Accessed November 15, 2021

Carayannis, E. G., & Campbell, D. F. J. (2009). "Mode 3" and "quadruple helix": Toward a 21st century fractal innovation ecosystem. *International Journal of Technology Management, 46*(3/4), 201–234. https://doi.org/10.1504/IJTM.2009.023374

Carayannis, E. G., & Campbell, D. F. J. (2010a). Triple helix, quadruple helix and quintuple helix and how do knowledge, innovation and the environment relate to each other? A proposed framework for a trans-disciplinary analysis of sustainable development and social ecology. *International Journal of Social Ecology and Sustainable Development, 1*(1), 45–69.

Carayannis, E. G., & Campbell, F. J. (2010b). Open innovation diplomacy and a 21st century fractal research, education and innovation (FREIE) ecosystem: Building on the quadruple and quintuple helix innovation concepts and the "mode 3" knowledge production system. *Journal of the Knowledge Economy*. https://doi.org/10.1007/s13132-011-0058

Carayannis, E. G., & Campbell, D. F. (2014). Developed democracies versus emerging autocracies: Arts, democracy, and innovation in Quadruple Helix innovation systems. *Journal of Innovation and Entrepreneurship, 3*, 12. https://doi.org/10.1186/s13731-014-0012-2

Carayannis, E. G., & Campbell, D. F. J. (2019). Innovation systems in conceptual evolution: Mode 3 knowledge production in quadruple and quintuple helix innovation systems. In *Smart quintuple helix innovation systems. Springer briefs in business*. Springer. https://doi.org/10.1007/978-3-030-01517-6_5

Carayannis, E. G., & Campbell, F. J. (2021). Democracy of climate and climate for democracy: The evolution of quadruple and quintuple helix innovation systems. *Journal of the Knowledge Economy, 12*, 2050–2082. https://doi.org/10.1007/s13132-021-00778-x

Carayannis, E. G., & Morawska-Jancelewicz, J. (2022). The futures of Europe: Society 5.0 and Industry 5.0 as driving forces of future universities. *Journal of the Knowledge Economy*. https://doi.org/10.1007/s13132-021-00854-2

Carayannis, E. G., & Morawska, J. (2023). Digital and green twins of Industry & Society 5.0. The role of universities. In E. G. Carayannis, E. Grigoroudis, I. A. Iftimie, & D. F. J. Campbell (Eds.), *Handbook of research on artificial intelligence, innovation and entrepreneurship*. Edward Elgar Publishing.

Caryannis, E. G., & Rakhmatullin, R. (2014). The quadruple/quintuple innovation helixes and smart specialisation strategies for sustainable and inclusive growth in Europe and beyond. *Journal of the Knowledge Economy, 5*(2), 212–239. https://doi.org/10.1007/s13132-014-0185-8

Carayannis, E. G., Ackidilli, G., & Ziemnowicz, C. (2020a). Creative destruction in international trade: Insights from the quadruple and quintuple innovation helix models. *Journal of the Knowledge Economy, 11*, 1489–1508. https://doi.org/10.1007/s13132-019-00599-z

Carayannis, E. G., Campbell, D. F. J., & Grigoroudis, E. (2021a). Democracy and the environment: How political freedom is linked with environmental sustainability. *Sustainability, 13*, 5522. https://doi.org/10.3390/su13105522

Carayannis, E. G., Campbell, D. F. J., & Grigoroudis, E. (2022). Helix trilogy: The triple, quadruple, and quintuple innovation helices from a theory, policy, and practice set of perspectives. *Journal of the Knowledge Economy, 13*, 2272–2301. https://doi.org/10.1007/s13132-021-00813-x

Carayannis, E. G., Christodoulou, K., Christodoulou, P., Chatzichristofis, S., & Zinonos, Z. (2021b). Known unknowns in an era of technological and viral disruptions—Implications for theory, policy, and practice. *Journal of the Knowledge Economy.* https://doi.org/10.1007/s13132-020-00719-0

Carayannis, E. G., Draper, J., & Bhaneja, B. (2020b). Towards fusion energy in the Industry 5.0 and Society 5.0 context: Call for a global commission for urgent action on fusion energy. *Journal of the Knowledge Economy*, 1891–1904. https://doi.org/10.1007/s13132-020-00695-5

Carayannis, E. G., Dezi, L., Greogri, G., & Calo, E. (2021c). Smart environments and techno-centric and human-centric innovations for Industry and Society 5.0: A quintuple helix innovation system view towards smart, sustainable, and inclusive solutions. *Journal of the Knowledge Economy.* https://doi.org/10.1007/s13132-021-00763-4.

Correia, A., & Reyes, I. (2020). *AI research and innovation: Europe paving its own way.* Working Paper, European Commission. https://doi.org/10.2777/264689

Churski, P., & Kaczmarek, P. (2022). Co-creation for developing the local climate emergency responses—Lessons from Poznań Agglomeration in Poland. In *International Conference LOCAL GOVERNANCE IN A TIME OF GLOBAL EMERGENCIES IGU Commission Geography of Governance Annual Conference—2022*, Mexico City, August 29–September 1, 2022. Researchgate. Accessed October 15, 2022

Costa, J., & Matias, J. C. (2020). Open Innovation 4.0 as an enhancer of sustainable innovation ecosystems. *Sustainability, 12*, 8112. https://doi.org/10.3390/su12198112

Deguchi, A., Hirai, Ch., Matsuoka, H., Nakano, T., Oshima, K., Tai M., & Tani, Sh. (2018). What is Society 5.0? In *Society 5.0: A people-centric super-smart society* (pp. 1–25). Hitachi-UTokyo Laboratory. https://doi.org/10.1007/978-981-15-2989-4. Accessed September 28, 2021

European Universities Association. (2021). *Universities without walls. A vision for 2030.* Report available on https://eua.eu. Accessed February 4, 2022

Fagiewicz, K., Churski, P., Herodowicz, T., Kaczmarek, P., Lupa, P., Morawska-Jancelewicz, J., & Mizgajski, A. (2021). Cocreation for climate change—Needs for actions to vitalize drivers and diminish barriers. *Weather, Climate and Society, 13*, 550–570. https://doi.org/10.1175/WCAS-D-20-0114.1

Fukuyama, M. (2018). Society 5.0: Aiming for a new human-centered society. *Japan Spotlight, 1*, 47–50.

Goddard, J., Hazelkorn, H., Kempton, L., & Vallance, P. (2016). *The civic university: The policy and management challenges.* Edward Elgar Publishing.

Grau, F. X., Escrigas, C., Goddard, J., Hall, B., Hazelkorn, E., & Tandon, R. (2017). Towards a socially responsible higher education institution: Balancing the global with the local, GUNI Report. Girona.

Hamaguchi, M. (2020). Imagine universities as prototyping places for social transformation, Michinari Hamaguchi. The future of universities. In B. Orazbayeva, A. Meerman, V. Galan Muros, & T. C. Plewa (Eds.), *Thoughtbook. Universities during times of crisis.* University Industry University Network.

Jørgensen, T. M., & Claeys-Kulik, A.-L. (2021). *Pathways to the future. A follow-up to "Universities without walls—A vision for 2030".* EUA. https://eua.eu/resources/publications/983:pathways-to-the-future.html. Accessed October 10, 2021

Owen, R. (2021). *Enabling open science and societal engagement in research.* Independent expert report. European Commission. https://doi.org/10.2777/057047

Potočan, V., Mulej, M., & Nedelko, Z. (2021). Society 5.0: Balancing of Industry 4.0, economic advancement and social problem. *Kybernetes, 50*(3), 794–811. https://doi.org/10.1108/K-12-2019-0858

Prainsack, B., Carayannis, E. G., & Campbell, D. F. J. (2012). Mode 3 knowledge production in quadruple helix innovation systems: 21st-century democracy, innovation, and entrepreneurship for development. *Minerva, 50*, 139–142. https://doi.org/10.1007/s11024-012-9194-6

Rego, B., Javantilal, S., Ferriera, J. J., & Carayannis, E. G. (2021). Digital transformation and strategic management: A systematic review of the literature. *Journal of the Knowledge Economy.*

Riviezzo, A., Napolitano, M. R., & Fusco, F. (2019). Along the pathway of university missions: A systematic literature review of performance indicators. In A. D. Daniel, A. A. C. Teixeira, & M. T. Preto, (Eds.), *Examining the role of entrepreneurial universities in regional development* (pp. 24–50). Universidade de Lisboa. https://doi.org/10.4018/978-1-7998-0174-0

Serpa, S., & Ferreira, C. M. (2019). Society 5.0 and sustainability digital innovations: A social process. *Journal of Organizational Culture, Communications and Conflicts, 23*(1).

Society 5.0 A People-Centric Super-Smart Society. (2018). Hitachi-UTokyo Laboratory. https://doi.org/10.1007/978-981-15-2989-4. Accessed September 28, 2021

Matsuoka, H., & Hirai, Ch. (2018). Habitat innovation. In *Society 5.0 a people-centric super-smart society* (pp. 25–43). Hitachi-UTokyo Laboratory. https://doi.org/10.1007/978-981-15-2989-4. Accessed September 28, 2021

Verhoef, L. A., Bossert, M., Newman, J., Ferraz, F., Robinson, Z. P., Agarwala, Y., Wolff, P. J., Jiranek, P., & Hellinga, C. (2020). Towards a learning system for university campuses as living labs for sustainability. In W. L. Filho, A. Lange Salvia, R. W. Pretorius, L. Londero Brandli, E. Manolas, F. Alves, U. Azeiteiro, J. Rogers, C. Shiel, & A. Do Paco (Eds.), *Universities as living labs for sustainable development supporting the implementation of the sustainable development goals.* https://doi.org/10.1007/978-3-030-15604-6

Zhang, J. J., Wang, F. Y., Wang, X., Xiong, G., Zhu, F., Lv, Y., Hou, J., Han, S., Yuan, Y., Lu, Q., & Lee, Y. (2018). Cyber-physical-social systems: The state of the art and perspectives. *IEEE Transactions on Computational Social Systems, 5*(3), 829–840. https://doi.org/10.1109/TCSS.2018.2861224

The Process of Selecting Influencers for Marketing Purposes in an Organisation

Tia Huttula◉ and Heikki Karjaluoto◉

Abstract Influencer marketing practices are growing on social media channels, while the usage of other mass-media channels is decreasing, prompting organisations to search for new tools with which to communicate efficiently with their target audiences. Influencers can affect purchase intentions if the audience identifies with them. For a successful collaboration with an influencer, an organisation needs to ensure that the brand fit is suitable, as the audience will become suspicious if the paid collaboration is too apparent. In addition, an organisation can ask an influencer to perform many roles during the collaboration. The objectives of this study, therefore, are to discover how organisations ensure brand fit with the influencer and to identify the roles fulfilled by the influencer in the organisation. To gain an understanding of the research topic, this qualitative research uses interviews with organisations, and interviews with media and influencer agencies located in Finland. Comparing the findings with previous research, two main implications were found. First, to ensure brand fit, the influencer's target audience is carefully checked to see if it matches the organisation's target audience; the values and content of the influencer are then checked in order to understand their character. The second implication is that the influencer's roles include (among other roles) those of content creator and protagonist. The theoretical implications suggested additional steps for the influencer selection process in order to ensure the brand fit is more accurate; there was also an implication that the extensive use of the influencer's roles would result in more effective outcomes.

Keywords Social media influencer · Influencer · Influencer marketing · Communication strategy · Brand fit

T. Huttula · H. Karjaluoto (✉)
School of Business and Economics, University of Jyväskylä, Jyväskylä, Finland
e-mail: heikki.karjaluoto@jyu.fi

C. F. Machado and J. P. Davim (eds.), *Industry 5.0*,
https://doi.org/10.1007/978-3-031-26232-6_2

1 Introduction

Social media is part of normal life and is gaining popularity every day, while the popularity of mass-media channels, such as print and television, is declining, prompting organisations to search for new methods for reaching their consumers (Bakker, 2018; Colliander & Erlandsson, 2015; Sundermann & Raabe, 2019). Organisations are increasingly starting to pay influencers to create content on behalf of the organisation and share it on the influencers' social media channels in collaboration with the organisation (Sundermann & Raabe, 2019). Regarding social media, consumers have reported that they trust the influencers they follow and the influencer reviews they find on social media (Kapitan & Silvera, 2016). Thus, the opinion of other people is an important factor when influencing human behaviour (Djafarova & Rushworth, 2016).

According to Borchers (2019), influencer marketing has become a mass phenomenon within the past few years, and according to Statista (2022), in the USA the influencer marketing industry has increased from \$1.7 billion in 2016 to \$16.4 billion in 2022.

Influencers' followers find the influencers to be trustworthy, credible, authentic and expert (Pöyry et al., 2019; Sundermann & Raabe, 2019). Influencers have significant numbers of followers that they can speak to and influence through their channels (Sundermann & Raabe, 2019). Furthermore, these characteristics have made organisations start to consider influencers as relevant intermediaries in their strategic communication.

With digital technologies, organisations have an easy, direct way to communicate with their customers (Bakker, 2018). Social media has become a popular communication platform mix, with communication taking place on various sites, including Facebook, Instagram and Twitter; these platforms host influencer marketing. Furthermore, Bakker (2018, p. 80) defined *influencer marketing* as 'a process in digital marketing where opinion leaders (influencers) are identified and then integrated into a brand's brand communication on social media platforms.' Organisations gain marketing and public relations value from collaboration with influencers (Borchers, 2019).

To reach their consumers, organisations can use influencers in many ways. These include taking on the roles of an intermediary (the influencer shares sponsored content), a brand content distributor, a creative content producer, an event documenter, a strategic counsellor and an event host (Borchers, 2019; Evans et al., 2017). Once, these posts were filled by different employees within the organisation; now, the influencer can fulfil these roles and enables new functions in strategic communication (Borchers, 2019). As a new form of communication, influencer marketing should help organisations reach their communication goals within the social media sphere (Bakker, 2018). Pöyry et al. (2019) studied influencer collaborations as components of marketing processes, and Sundermann and Raabe (2019) studied influencer communication from the perspective of brands, influencers and consumers.

Since then, strategic influencer communication has gained significant recognition in communication strategy research. However, research into this novel concept is uncommon, and there is a lack of understanding of the way in which influencers can be used efficiently to secure success for an organisation in the future (Enke & Borchers, 2019; Pöyry et al., 2019). Although the use of endorsers to promote an organisational message is not a new concept, the popularity of social media channels has changed organisations' approaches to the use of influencers (Pöyry et al., 2019).

In the literature of Sundermann and Raabe (2019), they found that 13% of all internet users and 50% of teenagers have purchased a product endorsed by an influencer; in addition, a survey found that 83% of 102 organisations already used or planned to use influencers and that 53% of the organisations had created a department to administer influencers (Sundermann & Raabe, 2019).

Influencers can speak the audience's language and are therefore seen as authentic; their sponsored content is not perceived as being as intrusive as organisational advertising (Bakker, 2018). The influencers' communication is seen as coming from a fellow social media user who speaks to and share content with other users using the same language as them. Moreover, when attaining trust, influencers have the potential to stand out from the clutter of ads and build meaningful relationships with consumers as trustworthiness is an important concept when considering the influencer's ability to persuade customers to buy a product or service and to change their attitudes (Bakker, 2018; Ohanian, 1990). Influencers form an important part of the purchasing decision journey as their main goal when collaborating with an organisation is to encourage a purchase (Bakker, 2018).

The influencers can share the brand image of the organisation with the audience and influence opinions. Therefore, it is important to choose an influencer with the right brand-fit (Bakker, 2018). To avoid raising audience suspicion about the authenticity of the influencer, the relationship between the product and the influencer should be logical; such audience suspicion on the authenticity of the collaboration could have a damaging effect on both the influencer and the organisation's brand image (Pöyry et al., 2019).

Additionally, previous research has found that the influencer has the ability to fulfil many roles in an organisation, depending on that organisation's goals for the collaboration, such as having a content creator, public persona and content distributor (Enke & Borchers, 2019). Organisational uses of these roles depend on the communication strategy's objectives and goals the role address (Enke & Borchers, 2019).

Audiences are resistant to traditional media, such as TV and print ads; they seek authentic, trustworthy information and thus turn to influencers (Bakker, 2018). Many organisations have therefore implemented influencer communication as part of their communication strategy, often in an unstructured way (Sundermann & Raabe, 2019). Thus, the brand fit of influencers and the roles they can fulfil in an organisation are relevant topics for research (Bakker, 2018; Enke & Borchers, 2019; Sundermann & Raabe, 2019).

This study aims to gain an understanding of the influencer selection process in an organisation and to discover how organisations use influencer communication in their communication strategy. A qualitative approach was chosen for conducting this research on the brand fit of an influencer and the roles of influencers in an organisation's communication strategy.

Therefore, the research questions for this study are as follows:

RQ1: How does an organisation ensure brand fit with an influencer?
RQ2: What roles does an influencer fulfil in an organisation's strategic communication?

This chapter includes five sections: the introduction, a literature review, a methodology section, a section on the research findings and a discussions section. References and an appendix can be found at the end.

This chapter follows the three stages of qualitative research: (1) explain the purpose and the concepts of the study with previous theory, (2) analyse and display the qualitative data and (3) discuss and present the implications of the findings (Malhotra et al., 2012).

2   The Literature Review

2.1   Brand Fit

The origin idea of brand fit is based on the match-up hypothesis, which examines the impact of different types of endorsers on a product (Till & Busler, 2000). The purpose is, therefore, to understand the fit between an endorser and a product (Till & Busler, 1998).

Furthermore, influencers speak the same language as their audience, and organisations are seeking storytellers who are intimate with their audience and can deliver their brand image and message in a trustworthy and authentic manner. Hence, organisations are not interested in influencers who would do no more than lend their name to a product (Hou, 2018). Based on the match-up hypothesis discussed by Kahle and Homer, the better the fit between the product and the endorser, the more effective the endorser will be. Suitable congruence between endorser and product leads to greater endorser plausibility than when the fit between them is less compatible.

Therefore, it is important to pick the right influencer for an organisation, ensuring that he or she has the right brand fit (i.e. the influencer's personality, brand and content fit the organisation's needs) and 'target audience-fit' (i.e. the influencer's target audience matches the organisation's target audience) (Bakker, 2018, p. 81). When an organisation is selecting an influencer, it is important to understand the characteristics that appeal to and influence the target audience (Bakker, 2018).

In order to gain brand awareness, the influencer should be well-known and trustworthy, characteristics that consolidate brand attitudes (Bakker, 2018). The influencer should have gained expert status among the audience, and the product category should be within his or her field of expertise (Bakker, 2018). To ensure that the organisation's strategic goals are achievable, the sponsored content must be aligned with the influencer's usual content and style (Pöyry et al., 2019). An inherent brand match between influencer and organisation leads to better results for both parties (Till & Busler, 1998). The audience perception of the influencer as being someone similar to them creates a peer-to-peer effect in their communication (Bakker, 2018). Also, if influencers are easily likeable and appealing, they can enhance brand attitude (Bakker, 2018). The influencer's power refers to an influencer's effectiveness in transforming a purchase intention into a purchase decision.

The relationship between the influencer and the product should be logical for the audience (Johnstone & Lindh, 2018; Keel & Nataraajan, 2012; Pöyry et al., 2019). Colliander and Erlandsson (2015) suggested that the influencer should carefully choose the organisation for product collaborations and consider the outcomes of sponsored content. A bad match of influencer and product can negatively affect brand image, decreasing the authenticity and credibility of the influencer's content (Pöyry et al., 2019). In addition, false and invalid statements about a product promoted in this context raise negative attitudes towards both the brand and the influencer (Djafarova & Rushworth, 2016).

For an influencer to be successful in promoting a product, organisations should consider some distinctive features, which should correspond with the organisation's goals (Bakker, 2018). The followers of the influencer make the influencer; the more followers, the bigger the possible reach is on social media channels (Bakker, 2018). However, in 2018, Neuendorf (as cited in Bakker, 2018) argued that influencers with smaller fan bases are more connected to the fans and have a better relationship with the target audience, so followers' growth rate and qualityscore are therefore the more important measures (Bakker, 2018, p. 83). *Growth rate* here refers to the growth of followers every month, and *qualityscore* refers to their engagement; these metrics help organisations to understand the followers and to discover follower overlaps in the different social media channels (Bakker, 2018).

Furthermore, a study by Deges (as cited by Bakker, 2018, p. 83) defined further features, the '4 R's of reach, relevance, resonance and reputation,' which are explained in more detail in Table 1. These metrics also allow organisations to measure if an influencer would be suitable for collaboration.

In conclusion, organisations should carefully choose the right influencer to fit the purpose of influencer marketing. The 4 R's offer guidelines for the selection process, help to identify important points for consideration during the process and ensure the best possible brand fit and target-audience fit (Bakker, 2018). To ensure the influencer's content stays aligned with the usual content, the brand match of the influencer and the product is mandatory since it is only with this match that the audience perceives the influencer's content as authentic and only then that the influencer is able to influence the audience.

Table 1 The 4 R's (Deges, 2018)

The 4 R's	Metric type	Description
Reach	Quantitative	The number of followers
Relevance	Qualitative	The fit of the influencer in regard to different segments that the organisation has defined: brand, target audience, content and personality
Resonance	Qualitative	The average interaction between the influencer and the audience measured by, for example, 'like follower rate' or 'comments per post'
Reputation	Qualitative	Is the influencer an expert in the field? How is the influencer's personality characterised, and is it compatible with the brand?

2.2 Strategic Communication

In social media, organisations can build brand image, engage with their audience and increase traffic to their online and offline stores. Marketing metrics—such as reach, click-through rates and sales—apply to social media. However, Macnamara (2018) argued that organisations find it difficult to prove engagement and the increase of brand image. Previous studies have shown that organisations rely on vanity metrics—such as reach, clicks and likes—when it comes to measuring communication's effectiveness on social media; even as social media is becoming more important and as analysing tools develop, communication attempts online are still assessed using meaningless measures (Macnamara, 2018). Because they lack proof of the impact of communication attempts on organisational goals, communication professionals face challenges when it comes to the evaluation of strategic communication (Macnamara, 2018).

To take steps towards the evaluation of communication strategy, evaluation models assist in understanding the logic used in strategic communication; Macnamara (2018) therefore introduced the integrated evaluation model, which integrates communication features with the two-way flows connecting stakeholders and the public. The model recognises the overlap of and the need to rely on the communication evaluation stages (inputs, activities, outputs, outcomes and impact). To analyse the two-way flows of this model, the outputs flow from the organisation to the stakeholders and the public; the outcomes and impact subsequently return to the organisation (Macnamara, 2018). Hence, Macnamara (2018, p. 193) stated that communication evaluation models reveal '*what* is intended to be done to *whom* and *whose* interests are served' in the communication strategy process.

Furthermore, influencer marketing communication depicts the traditional organisational communication strategies as organisations need to establish trust with consumers; in addition, influencer marketing is an effective tool for reaching fragmented audiences (Bakker, 2018). For organisations, influencer collaborations offer the possibility to increase trust when the brand fit is suitable; organisational communication, therefore, could be seen more as authentic communication than as advertising (Bakker, 2018).

2.3 Influencers in Marketing Strategy

Because many organisations identify influencers as intermediaries in making contact with hard-to-reach stakeholders through influencers' channels, research into strategic communication has adopted strategic influencer marketing as a major topic (Enke & Borchers, 2019). Furthermore, Enke and Borchers (2019) stated that strategic communication research is not interested in influencers as influencers but wants to understand their role in communication strategy. They described influencers as 'secondary stakeholders' with the ability to influence 'primary stakeholders; influencers can also act as primary stakeholders when they are creating content for organisations (Enke & Borchers, 2019, p. 263). Influencers also offer possibilities to shift brand images through strategic communication, especially when the organisation has a long-term (rather than a one-off) collaboration with the influencer (Borchers, 2019).

Furthermore, Enke and Borchers (2019) suggested that organisations should consider the function which is fulfilled by the influencer in the processes of communication and organisational value creation. They also mentioned that communication and measurement models allow organisations to systematically consider the organisational objectives in relation to the collaboration with influencers and the different stages affected by the collaboration in communication processes (Enke & Borchers, 2019).

The first stage of the communication process, the input stage, relates to the influencer and the communication strategy; the influencer could provide (a) material resources (such as technical equipment, manufacturing resources, authoritative competence and relationships) and (b) internal or external organisational resources.

Enke and Borchers (2019, p. 263) identified the external resources that the influencer could provide as 'seven external resources that organisations try to harness by cooperating with influencers: content production competences, content distribution competences, interaction competences, a public persona, a significant number of relevant relationships, a specific relationship quality, and the ability to influence.' Table 2 visualises the roles and outputs the influencer then can provide for the organisation from these resources they have.

Below the influencer roles are described:

Content creator: This refers to the content creation role of influencers, which can be executed individually by the influencer or in collaboration with the organisation. The organisation relies on the influencer's content creation role and acquires the content for the organisation's channels (Enke & Borchers, 2019).

Table 2 An influencer's resources in strategic communication (Enke & Borchers, 2019)

Roles	Output
Content creator	Content
Multiplicator	Reach
Moderator	Interaction
Protagonist	Personalisation
–	Relevant contacts
–	Peer effect
–	Influence

Multiplicator: In this role, the influencer distributes the organisational messages on his or her platform (Enke & Borchers, 2019).

Moderator: In this role, the influencer can engage with topics that are relevant for the organisation, for example, he or she can interact in public discussions and influencer gatherings (Enke & Borchers, 2019).

Protagonist: In this role, the influencer, a main character for the organisation, can perform at events as a host, expert or discussant (Enke & Borchers, 2019).

Thus, the organisation chooses the number and nature of the roles that are included in the collaboration; they may decide on the influencer having one role or multiple roles (Enke & Borchers, 2019). Other inputs that do not involve direct activity enhance the effectiveness of the influencer and should not be ignored (Enke & Borchers, 2019).

The outputs resulting from the role of the influencer in a collaboration follow:

Content: This can consist of text, pictures and videos, depending on the influencer's competence and agreed role as a content creator.

Reach: This involves the views of content in which the influencer shares the organisation's content.

Interaction: This includes, for example, likes, shares and comments in social media channels. The influencer's role here is to initiate and direct conversations as well as to keep them going (Enke & Borchers, 2019).

Personalisation: The influencer personalises the message; organisations benefit from the authentic content. Personalisation can have an effect on brand image and organisations are advised to consider the content style of the influencer before starting the collaboration (Enke & Borchers, 2019).

Relevant contact: The organisation can attempt to gain an audience who will follow it directly rather than only reaching the audience through the influencer (Enke & Borchers, 2019).

Peer effect: The influencer's authenticity and credibility can lead to a peer-to-peer effect between the influencer and the audience; in strategic communication, this could be instrumental in achieving the objectives of the collaboration (Enke & Borchers, 2019).

Influence: The main goal of strategic influencer communication is to influence the target audience in such a way that the objectives of the collaboration with the influencer are accomplished (Enke & Borchers, 2019).

The outcome for a collaboration is effected by combination of the roles and outputs which are selected to support the collaboration objectives (Enke & Borchers, 2019). The objectives for collaboration could be changes in brand awareness, attitudes or behaviour. Enke and Borchers (2019) mentioned that the mix of these different roles and outputs results in effective outcomes on strategic influencer communication. For the same reason, Borchers (2019) suggested using influencers on various platforms and with different content formats, such as text, pictures, videos and live streaming.

2.4 The Framework and Conclusions of the Theory

The theoretical model of this research has been formed to establish the basic knowledge of the subject, describing previous and more recent studies in the field of influencer marketing, as well as describing the influencer's part in strategic communication.

Influencersbeing experts, in a particular field in the minds of their friends, families and acquaintances—can influence others on their social media channels (Freberg et al., 2011).

Influencers contribute to an important part of the purchase journey; their main role in collaboration with organisations is to encourage audiences to make a purchase decision (Bakker, 2018). Audiences are more responsive to the sponsored content from influencers than to ads from organisations; although they understand that influencers promote products, they trust them not to abuse their trust by giving false reviews (Djafarova & Rushworth, 2016).

The fit between the product and the influencer needs to be logical if the audience is to accept the message. For this reason, match-up hypothesis studies discuss the characteristics that are important for the fit, such as attractiveness, relatedness, similarity and consistency (Till & Busler, 2000). As the fit was found to be necessary, Bakker (2018) presented Degas's 4 R's that can be checked in order to optimise the fit when selecting influencers.

However, interaction with an influencer is perceived as more trustworthy than organisational communication, so organisations are increasingly using influencer communication as part of their strategic communication (Sundermann & Raabe,

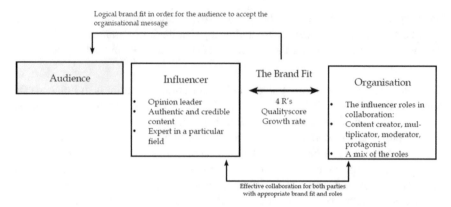

Fig. 1 The framework of the literature review

2019). Enke and Borchers (2019) described the influencer's roles in strategic communication as being those roles that the influencer can execute during the collaboration and those that could be expected from the roles in the collaboration. Figure 1 presents the framework of the theory.

3   The Method

The goal of this research is to understand how organisations ensure brand fit with an influencer. A qualitative method is used in this study to elaborate on the personal (organisational) experiences of a phenomenon and gain an in-depth understanding of the current situation regarding the relationship between influencer and organisations, and of the roles that the influencer can fulfil in strategic communication. The techniques of qualitative research are unstructured and not predefined; the research might be modified during the data collection if new attention points come up, and the issues explored can change as the project develops (Malhotra et al., 2012).

Interviews are chosen as the research method for this study because they allow the interviewer to gain an in-depth understanding of each participant's experience and beliefs regarding the topic. Four organisations from the retail field and two agencies were interviewed to gain knowledge. The interviewees from the organisations were marketing managers and the head of influencer marketing; one media agency and one influencer agency took part, to ensure understanding of the topic and their processes.

Interviews enable researchers to gather vast amounts of information (Adams et al., 2014; Malhotra et al., 2012). Face-to-face semi-structured interviews were conducted at each interviewee's choice of location; all of the interviewees were located in the Uusimaa region of Finland. Each interviewee was contacted by phone and email, allowing the researcher to describe the research and the interviewee's part in it.

The goal of the semi-structured interviews was to gain meaningful answers that would help attain the research objective (to understand the relationship between influencers and organisations) by interviewing Finnish influencer marketing experts about current relationships and Finnish organisations about their use of influencers as part of their communication strategies. The interviewees were sent an information package telling them about the time they should allocate for the interview (1 h), and the themes and topics of the interview; the package also contained preliminary questions (see the Appendix) for interviewees to think about before the interview took place.

All the interview participants were knowledgeable about influencer communication in their organisations, and each organisation included in the research had collaborated previously with an influencer before the interview.

All the interviews were conducted in Finnish, the interviewees' working language; they preferred using it to ensure they could express themselves correctly. Each interview was transcribed directly into English from the recordings and the transcription was then sent to the interviewee to check that the English vocabulary and expressions approximated their vocabulary and style of speaking. At this stage, any content that could compromise the interviewee's anonymity was changed to allow the interviewer to retain his or her anonymity. For example, an influencer's name was replaced with '[a/the influencer]'. The three main topics in the interview were the organisation's use of influencers and prior experience of using influencers, the relationship between the organisation and influencer, and the organisation's communication strategy. Each interview, which lasted between 19 and 45 min, covered all the topics. Data anonymity was ensured during transcription; the interview participants approved the transcripts before the analysis was conducted. Only six interviews were conducted as the data saturated rapidly, with responses quickly starting to resemble each other.

Content analysis is used to describe the content gathered from an interview and analyse it systematically and objectively (Adams et al., 2014). The aim of this method is to summarise and order the data to enable conclusions to be drawn from the data (Malhotra et al., 2012). The characteristics of the analysis could be key words, themes, characters, topics or space and time (Malhotra et al., 2012). Some themes, based on the theory and planned topic areas, can be expected to arise from the interviews, but others can arise from the analysis of the data (Adams et al., 2014).

The analysis started with a data reduction. The data were then entered in Excel to visualise the responses to the specified themes and to place other findings in one table, looking for other unexpected findings that might occur. Furthermore, the interviewees were divided into two categories: organisations and agencies. In the analysis, each was given a code consisting of letters (org for an organisation, agc for an agency) and a number, for example, org1 or agc1. Table 3 presents background information about each interviewee's job title and age, and the length of his or her interview.

The data display step takes the analysis forward sensibly, as it presents the data in a visual format, with the concepts and relationships all in one location (Caudle, 2004). When data are visually presented, new relationships and explanations can be found that were not expected in the research proposition; critical thinking is also easier to conduct (Caudle, 2004). At this stage, illustrations of the answers to the interview questions could be presented; for example, if a theoretical approach is

Table 3 The participants in the interviews

Organisation/agency identification code	Job title	Age	Duration of the interview (in minutes)
org1	Content lead	28	29
org2	CEO	50	25
org3	Influencer manager	32	19
org4	Marketing manager	34	29
agc1	Client manager	30	26
agc2	Creative director	29	45

taken, an illustration of the theory could be presented as a measure and the responses placed accordingly (Adams et al., 2014). The final part of the analysis is to interpret the data and form a conclusion; this could include, for example, a review of the notes, a comparison of perceptions, and a search for patterns and connections that explain a phenomenon (Adams et al., 2014).

4   Findings

4.1   Experiences with Social Media Influencers

To get an understanding of how much the interviewee had to say about the topic based on his or her experience, the first question was about previous his or her experiences with influencers in their organisation. For this section of the interview, most of the interviewees brought up details of the selection process, identifying the important aspects of choosing the best influencer, collaboration goals and the channels used. In addition, agc1 and agc2 gave insights into how the influencers had been used previously and, from their perspective, what organisations ask for when planning marketing or campaigns. It must be mentioned that all the interviewees stated that influencer marketing was a new thing for them, having been implemented in their organisations within the past five years.

The agencies stated that influencer marketing was gaining popularity. Before, collaborations were used for product launches and rebranding when there was a need for a huge audience. Both agencies claimed that there is now a demand from organisations for brand ambassadors and long-term collaborations. This claim was supported by the organisations' interviewees as all of them preferred and (were aiming to find) long-term collaborations.

Compared with other marketing attempts, influencer marketing was seen by both agencies as cost-efficient because organisations could make product exchanges for the collaboration rather than having to pay for it. The reach, content and brand image achieved through the influencer had made the collaboration worthwhile.

In the organisations, the influencer acquisition process was different from that of the agencies, who explained that they were briefed by an organisation, and kept the details (especially the target audience and the goal) in mind when starting to search for the influencers. The organisations received many collaboration proposals through email and Instagram Direct (even receiving hundreds of contacts weekly) but said that they had only started collaborations with a couple of the people who had made contact. Other ways to acquire influencers were by contacting media or influencer agencies (org1 was doing this) and searching for them on social media; influencers were also recommending each other to organisations. The organisations—org2, org3 and org4—were either searching for influencers themselves or being contacted by influencers.

4.1.1 Selecting the Right Influencer

The most important thing for the organisations when selecting the influencer was the target audience. Everyone highlighted the need to identify the influencer's target audience, with the influencer. They emphasised that they must know who they are addressing through the influencer; the agency interviewees also mentioned this, saying that they needed to know the target audience of the organisation if they were to find a suitable influencer. The interviewee for agc2 mentioned that her agency checks the location, age and gender—even the educational level—of the influencer's target audience. The same agency interviewee also wanted to mention here that the subjective effect needs to be taken into account when selecting the influencer. The person in charge of influencer selection might have a social media crush and so want to go with that particular influencer even if better influencer options existed.

In addition to the target audience, organisations found that a mix of influencers in different channels and the length of the collaboration were important. Explaining this, they said that they had found that a wide range of channels offered possibilities to reach different target audiences and that the collaboration would provide different content for them to use in other marketing channels. In addition, the mix of long-term and one-off collaborations provided variety for the organisations' target audience.

In the influencer selection process, the second important point that came up in all the interviews was that the influencer's values must meet the organisation's values and that those values need to be visible in the channel content. This is important when the brand image is built, and the organisations want the influencer to represent the brand image they have built. The interviewee for org1 mentioned that the organisation wanted to reshape the brand image through influencers in order to embed their desired brand image in the target audience; as an influencer was needed to do that, the values of the organisation and the influencer needed to match. Furthermore, the interviewee for agc1 supported the interviewee for org1's argument that the brand image could be directed to what is desired by the organisations with long-term influencer collaboration and an appropriate brand that fitted with the organisational values.

Furthermore, the interviewees for org1 and org4 both mentioned that the natural fit with the influencer's content was one of their biggest conditions for starting a collaboration, and if an influencer was already using their products, that was a positive addition.

On the other hand, the interviewee for agc2 pointed out that when a new product is launched, the match of the product and target audience cannot be proven before the results are compared with the collaboration goals, upon which the goals can be seen to have been met or not.

The influencer's audience was matched with the target audience as well as the influencer's values were matched with the organisation's values. Moreover, the general content of the influencer should fall within the field of the business, and there should be a theme to the content, for example, the influencer's lifestyle or expertise in a certain field. The influencer should fit with the brand image and elevate the brand during the collaboration. The interviewee for org4 mentioned that the organisation wanted the influencer to be brand loyal in order to make the collaboration natural and authentic.

4.1.2   The Relationship of the Organisation and Influencer

A relationship with the influencer was important for the organisations as it tied the influencer to the brand. The influencer should be easy to work with, and the relationship should be mutual, each wanting to work with the other and engage during the collaboration.

It was mentioned by organisations and agencies that the influencer should have an emotional relationship to the brand and that, for example, the products would be visible in his or her everyday life and not just in the one paid post. Relationship maintenance, especially in a long-term relationship, is important. To maintain the relationship, the interviewee for org1 had decided that the products should not be sent all at once but to be sent in smaller numbers over the time of the collaboration; this allowed the organisation to check on and update their level of satisfaction with the influencer's progress. On their social media channels, organisations shared all the posts about the collaboration that the influencer posted; this showed the influencer that the collaboration was important to the organisation.

Most organisations made a contract between themselves and the influencer; the agencies mentioned that the contracts could be formal or simply email agreements. Both the interviewees for org1 and org2 mentioned that they made contracts with long-term influencers, while the interviewee for org3 had an influencer brief that was to be followed during the collaboration; this brief mentioned, for example, restricted topics involving their products, such as the use of alcohol or drugs. The interviewee for org4 had only made spoken contracts with the current influencers, but felt that they did not need formal contracts because the agreement was so clear to both sides.

However, many things were mentioned that should be agreed upon with the influencer before the collaboration. Most of the organisations wanted to agree on the number of posts to be published, the times of publication and the type of content.

They also needed to formalise whether the influencer was to provide content for the organisation and needed to ensure that the influencer would not collaborate simultaneously with competitors. The interviewee for agc1 mentioned that if the influencer created content, the copyright would need to be carefully agreed upon before the collaboration to avoid misunderstandings and conflicts.

Such a contract could include a non-compete clause; the agency's job was to check for compliance. The interviewee for agc2 said that companies had even requested that the influencer should not have worked with a competitor in the past 18 months.

To support this, the interviewee for agc1 mentioned that they had had cases where a company refused to work with an influencer who had worked previously with a competitor. Moreover, all the organisations mentioned that they did not want competitors to be working simultaneously with the influencer; they considered this during the se-lection stage, asking whether the influencer had previously worked with two competing brands simultaneously.

Both the interviewees for org1 and agc1 argued that if influencer had worked with two competing brands simultaneously, organisations might consider: Do the influencers work with anyone they can get, or do they take good care of their brands? Other collaborations were also assessed; there should not be too many simultaneous collaborations lest the organisation's collaboration be lost among them. Otherwise, other collaborations were acceptable, as everyone under-stood that the job of an influencer is to collaborate with companies.

The interviewee for agc1 said that the companies were searching for 'normal people' who are easy to relate to and approach. They, therefore, used micro-influencers (defined by the interviewee for agc1 as influencers with fewer than 5000 followers on Instagram) to achieve unique collaborations.

The interviewee for agc2 agreed on using micro-influencers to achieve unique collaborations, but also said that they might be inexperienced collaborators and that they would sell too directly to the audience, thus compromising the authenticity of the collaboration. Both the interviewees for org1 and org3 mentioned that they were working with micro-influencers, org1 because of the overwhelming number of collaborations with popular influencer accounts and org3 due to their field of business. In addition, both of these interviewees mentioned that micro-influencer collaborations are more authentic because micro-influencers are usually already using their products.

As mentioned previously, all the organisations preferred long-term collaborations since they were perceived as trustworthy and credible. However, org3 was only doing one-offs at the time of the interview but had started to search for long-term collaborations or brand ambassadors. Despite that, org3 was having difficulties finding these long-term influencers because customer lifespans were short in their field; this applied to the influencers too.

The interviewee for org1 said that long-term collaboration was more reliable for them and gave them the ability to build a relationship with the audience. They used long-term collaborations for the main topics of the year and the more cost-effective one-offs for specific product campaigns. The interviewee for org2 mentioned that they used one-offs alongside long-term collaborations, maintaining interest by introducing a new face for a specific campaign.

The agencies supported the organisations' arguments, saying that long-term collaborations were credible, trustworthy and authentic, suitable for raising brand awareness and building customer relationships; the brand image could be moulded to meet the company's required direction. On the other hand, one-offs are tactical and campaign-specific or can be product launch collaborations.

4.1.3   The Goal for Influencer Collaborations

The goal of influencer collaborations, mentioned in all the interviews, was to gain brand awareness. In addition, although different key performance indicators (KPIs) were set for different collaborations, the main goal was always to gain brand awareness. The secondary goal was to increase sales.

Sales were easier to follow than the brand awareness; changes (or a lack of changes) in sales after a collaboration could be implied to have resulted from the collaboration, especially if it only involved one or a few products.

Other goals, mentioned several times, were to gain access to the channels the target audience uses, to gain more visibility for certain services or products, to highlight the organisation's values and product lines, to reach an audience when the organisation is topical for the audience and to gain new loyal customers. The interviewees said that many of these goals could be achieved through influencer collaborations; it was also stated that the goals needed to be clear at the start of the collaboration to enable an understanding of its success.

The organisations interviewees were asked how they would describe a successful collaboration. For the interviewee org1, a successful collaboration was one that had reached all the KPIs set for it. Brand awareness would have increased, and sales would have increased overall or the sales were measured with a promocode uses. At the end of the collaboration, the organisation would be happy with the influencer and possibly continue the collaboration later. The interviewees for org3 and org4 also mentioned the increase in product sales and brand awareness.

The interviewee for org2 thought more directly about increased sales and, on the side, brand awareness, also saying that the organisation would like to hear that people were starting to talk about the brand. Summing up, the interviewee for org2 highlighted that the collaboration should be easy going for both influencer and organisation.

The interviewee for org4 also said that the organisation would like positive feedback about the collaboration from consumers and hoped that the consumer would come to ask about the product made visible by the collaboration.

The agencies paralleled the organisations' responses. They mentioned good numbers, followers' interest and engagement in the collaboration, and sales, but above those, they highlighted the importance of communication between the company, influencer and agency, as well as having a positive feeling about the collaboration.

4.2   Influencer Functions in Strategic Communication

The roles of the influencer in the organisations were not as versatile as they could be. Agencies mentioned that influencers could be used as content creators for the company and for themselves; other possible roles include event host, photographer, meet-and-greet event host, brand or company protagonist, workshop expert or even campaign planner and concept creator. However, the organisations were not using influencers as broadly as that.

The influencers were used as content creators for their own channels; only the interviewees for org1 and org3 were asking for content for their organisations' use. The interviewee for org1 had also used influencers for meet-and-greet events and the interviewee for org2 had used an influencer for modelling in the organisation's photoshoot. For the future, all the organisations mentioned that they would like to hold meet-and-greet events, increase the use of influencer content in their marketing and have the influencers come to different events. They also said that the use of influencer marketing had not affected their organisation's internal roles noticeably; for example, the influencer had not become the only content creator for a campaign.

4.2.1   The Influencer Communication Strategy of the Organisation

Influencer marketing differed in each organisation's communication strategy. Influencer marketing was already implemented in org1's marketing and communication strategy, and was in its own section. The organisation's goal for the future was to use it in omnichannel communication to make it more visible and highlighted in other communications, even in offline channels. The organisation was currently sharing influencer posts on their own social media.

The second organisation, org2, based its marketing and communication strategy on influencer marketing and had planned its next year's expo event with an influencer. It hoped to develop an organisational strategy to get more influencer content onto its channels in the future; currently, it was only sharing its content or running photoshoots with the influencer to obtain content.

Furthermore, org3 did not plan collaborations ahead of time at all. The current strategy was that the organisation would work with an influencer if a suitable one was found or if one contacted them; from there, it only shared the posts the influencer had published on his or her social media. The plan for the future was to clarify the process of acquiring influencers and to start implementing them in the strategy for future campaigns or to find a brand ambassador.

The last interviewee, from org4, mentioned influencer marketing as a unique style for the organisation in Finland, making up a tiny, separate part of its marketing. The organisation had plans for influencer marketing to support its other marketing.

The planned future for org4 was to increase influencer support for other marketing plans, and even to amend the official content with the influencer content, thus improving the marketing fit in the Finnish market.

All in all, influencer marketing was considered to be separate from the marketing and communication strategy. However, all the organisations planned to implement more influencer marketing, making it visible and planned in their omnichannel strategies.

Agencies said that they planned influencer marketing to match the goals of the whole marketing strategy of a company, but also spoke of it as an independent operation. They also said that when a campaign or strategy was planned, possible collaborations were planned at the same time. In addition, they checked, for example, whether the influencer content could also be used in some other channel. The interviewee for agc1 mentioned that the agency had facilitated some omnichannel collaborations in which the influencer was visible in Google or Facebook ads, but that there was space for development in order to make influencer marketing more functional. On the contrary, the interviewee for agc2 argued that influencer marketing would not be mixed with other marketing in the future.

The agencies agreed that influencer marketing formed its part of the marketing and communication strategy, but unlike the organisations, they believed that it would not merge significantly with other marketing and communication strategies.

The measurement of marketing and communication—that is, using knowledge of the past as a guide for strategy development—is a big part of strategic planning. This is important but challenging for influencer marketing; all the organisations had difficulties with this, mentioning that they had trouble knowing what collaboration results to measure.

The measures used to follow the collaborations come from social media—such as likes, comments, reach and impressions—and the agencies provide social media numbers too. They mentioned that they could dig deeper and, for example, report the emotions raised by the collaboration. Sales provided another measure, but the agencies mentioned that that metric could not be used for every collaboration because the goal might be to gain brand awareness.

4.2.2 The Future Vision of the Influencer Marketing

In the interviews, many comments were made about the future and the direction in which the interviewees saw the trend of influencer marketing going. They believed it to be a growing field and were planning to invest in it as soon as next year. The interviewee for agc1 argued that influencers were becoming more professional, and companies were taking them more seriously. Organisations could, therefore, be increasing their investment in influencer marketing and were interested in seeing where the trend is going.

They also predicted that the trend would shift towards influencers working with the companies they really want to work with; influencers were stricter with the association of their brand with a company. For example, influencers wanted to test the products and ensure their quality before they collaborated. Interestingly, the inter-viewee for agc2 mentioned that even the influencers were starting to be more exacting with their brands; the agency interviewee had never heard of an influencer asking about the other influencers who were working with the company simultaneously.

4.3   A Summary of the Research Findings

Influencer marketing is becoming more popular and structured in organisations as it is a cost-effective way to reach target audiences that were not previously reach-able through the organisation's channels. For many, influencer marketing was still a new phenomenon, and there were concerns about ensuring its profitability and that the target audience had really been reached through the influencer. The influencer acquisition process had different practices, such as an organisational search for the influencer, an influencer approaching the organisation via email or Instagram Direct messages, or an organisational approach, using media and influencer agencies.

The organisations had structured ways to select and ensure brand fit with the right influencers as brand fit was important to every organisation. First, they checked the influencer's audience to ensure the audience fitted with their target audience—it was important to know those whom they would be addressing. The length of the relationship with the influencer and the channel were considered to vary the target audience reached. The second important point to check was the influencer's values and his or her presence in the content as an organisation's brand image could be affected and moulded by the influencer given a sufficiently long collaboration. For an authentic collaboration, therefore, the influencer and the organisation needed a natural fit.

The relationship between the influencer and the organisation should be natural and mutual, each wanting to work with the other. The influencer having an emotional tie to the product would be the goal, with the influencer using the product outside the collaboration. Organisations were strict about influencers' working with competi-tors, but other collaborations were acceptable as long as there were not so many that the organisation's collaboration would become insignificant for the influencer, risking the loss of the collaboration among other sponsored content. As the influ-encer field became more crowded, micro-influencers were gaining popularity. These micro-influencers were described as being more approachable people and, since they usually already used the organisation's product, the collaboration was more authentic than with bigger influencers. Furthermore, long-term collaborations were favoured, being perceived as more trustworthy, credible and positive for building customer relationships.

The goals for influencer collaborations were similar to other marketing goals. The first goal for all collaborations was to gain brand awareness, the second to gain sales. However, it was hard for organisations to measure the brand awareness; the increase of overall sales was easier to detect, and especially with collaborations for one specific product, increased sales of that product were easier to attribute to the collaboration. Other goals mentioned were access to the target audience, a highlight for some specific service, consumer enquiries and the acquisition of new loyal customers. Many goals were mentioned, but it was most important that the goals should be clear to both parties at the outset of each collaboration.

Among the influencers' channels, Instagram was the most popular for collaborations because of its visuality and text option. YouTube was also popular for targeting a younger audience and obtaining video for the organisations to use in other marketing material. Blogs were still used in some cases, especially when text content, such as educational content, needed to be available, but blog reader numbers are declining. Predictions were made about TikTok and podcast collaborations coming to Finland in the near future, and some of the organisations were, therefore, considering options for those too.

An influencer could fulfil many roles for the organisation, such as content creator, event host, protagonist and customer workshop expert. The influencer roles in the organisational collaborations studied were usually limited to content creation only. Influencers were creating content for their own channels; only in a few cases were they creating content for the organisation's own use in other marketing channels. Other roles used were a event host role at a meet-and-greet event and one modelling role in a photoshoot. For the future, organisations had planned more roles for influencersfor example, host roles at meet-and-greet events—and more content creation for other channels.

Influencer implementation into the overall communication strategy was seen as a separate part of the strategy, but the goal for organisations was to increase the use of influencers in their omnichannel strategy. Agencies, however, did not see that influencer marketing would merge with other marketing attempts. The importance of measuring collaborations, to know how they worked, was appreciated. Organisations were finding some measurements difficult as brand awareness is difficult to detect and, therefore, to measure. Sales, on the other hand, are easy to measure; increases and decreases can be directly linked to any collaboration that concentrates on a given product. Influencers can also provide the organisation with channel statistics, such as statistics on likes, comments and reach.

The organisations were going to increase investment in influencer marketing in the coming year as the trend for influencer marketing was increasing. Influencers are becoming more professional and starting to take care of their brands; therefore, they select collaborations more carefully. The organisations were interested in seeing the direction of the trend.

Table 4 illustrates this study's most important findings: ensuring band fit and influencer roles; from these, it is easy to continue onto their theoretical implications.

5   Discussion

This chapter offers two significant contributions: it highlights the importance of brand fit and the influencer's functions in an organisational communication strategy.

Table 4 The conclusions from the findings

Ensuring brand fit
Target audience check
The values of the influencer
Is there a natural fit with the influencer's content?
Is the theme of the content in the field of business?
Influencer roles
Content creator for the influencer's channels
Content creator for an organisation
Model in a photoshoot
Meet-and-greet event host

Regarding the first contribution, the brand fit process has many similarities to the model introduced by Deges (2018). He suggested using the 4 R's for selecting influencers. *Reach*, refers the number of followersthe first step for organisations to consider—was seen to be applied, as was the second R, *relevance*, which was applied because the organisations considered the target audience. However, the influencers reach in regard to the target audience was found to be more important to the organisations than the number of the influencer's followers.

Djafarova and Rushworth (2016) mentioned that an influencer was found credible when the match between the endorsed product and the influencer's source credibility characteristics was appropriate. Thus, the influencer's characteristics met the expectations of the audience—for example, the expectation that the influencer was an expert in the field of the product category; interestingly, this was not considered by the organisations in the findings. Also, an appropriate fit between the influencer and the product was important for the reception of positive brand attitudes, and such collaborations were seen as less intrusive than organisational advertising by the audience (Bakker, 2018). Furthermore, the brand fit between the product and the influencer needed be suitable in order to reach the best results from the collaboration; Till and Busler (2000, p. 578) mentioned that 'the effectiveness of the endorser varies by product'. Thus, the findings imply that the organisations were effectively trying to ensure the brand fit by carefully checking the values, content and target audience.

However, the third and fourth R's of Deges's (2018) model were not visible in the findings. *Resonance* and *reputation*s were not found, unlike the first two R's. Interestingly, the interaction between the influencer and the audience was not something the organisation would check. Reputation was considered mildly, under the heading of *personality characteristics*, for example, but expertise was not considered in the selection process.

In addition, for influencer selection, Bakker (2018) suggested evaluating the *qualityscore* (the engagement rate with the audience) and the *growth rate* of the followers. These measures help organisations to better understand the audience relationship with the influencer.

Furthermore, Borchers (2019) suggested that long-term collaborations enabled the influencer to affect brand image and increase image transfer; this applies to the findings as the organisations preferred long-term collaboration when their goal was to gain or create brand awareness, while one-offs were used for tactical collaborations when, for example, a product needed increased sales. It was also claimed that long-term collaborations were more trusted and credible, but Enke and Borchers (2019) suggested more research into long-term relationships between influencers and organisations. Hence, it could be interpreted that organisations were working with influencers in long-term relationships, although there was no certainty about the long-term effect on the brand. The long-term effect of an influencer collaboration should be further studied.

The first research question was, 'How does an organisation ensure a brand fit with an influencer?' The organisations ensured brand fit by carefully checking the influencer's values and content on the channels, as well as by checking that the influencer's audience matched the organisational target audience. They searched for a natural fit with the influencer's content and ensured that the influencer's content fitted their field of business. Further, the theory by Bakker (2018) suggested that the influencer's relationship with the audience should be checked, including checking factors such as the qualityscore, growth rate and interaction with the audience. Brand fit was found to be a significant factor for the influencer's credibility, trustworthiness and authenticity, all characteristics that influence the audience's purchase intentions.

In more detail, in organisations' influencer selecting process, four important steps could be followed when selecting an influencer with suitable brand and target-audience fits with the organisation.

First, the organisation is advised to check the size of the influencer's audience (the number of followers) and their characteristics, such as their age, gender and location distribution. The audience should fit the target audience of the organisation or of the campaign the influencer is to undertake. This first step helps to understand who the people are that the organisation will address through the influencer in question.

Second, the interaction between the audience and the influencer should be assessed in order to understand how the audience views the influencer and whether their relationship is engaging; this will indicate the influencer's ability to influence the audience. The interaction could be measured the with qualityscore, an average of the like follower rate and the comments per post, and with the growth rate of the followers.

Third, the reputation of the influencer should be considered. How does the audience perceive the influencer's content? Do they see the influencer as an expert in the field of the business of the organisation? What values does the influencer have, and are they visible in the influencer content? In order to explain the match between the content and the product or brand in question, their characteristics should be visible. This is particularly important because the influencer's values can affect the brand image of the organisation; the match—why the influencer is endorsing a certain product—should be obvious to the audience.

The last step involves the influencer's channel, which is connected to the target audience because different channels have different customer bases. The channels also affect the type of content created by the influencer. The content depends on the goal and the target audience: if video content is needed, the influencer should already be producing video content for YouTube, for example, and if text is needed, a blog writer could be ideal for the purpose.

During the selection process, these steps help to ensure a fit with the influencer's target audience and a fit between the influencer and the product. However, it is important to note that the length of the relationship was found to have an effect on the trustworthiness of the influencer as long-term relationships were perceived as more authentic and trustworthy and perceived to have the ability to change brand image. Therefore, organisations aiming to gain brand awareness should try to find influencers with whom they could work for longer periods; brand awareness could be compromised if the authenticity of the collaboration is questioned. On the other hand, authentic one-offs could be used to increase sales of a single product or service.

The second contribution relates to the strategic perspective; influencer marketing is becoming part of strategic communication as influencers are intermediaries in reaching target audiences that would not otherwise be reachable (Enke & Borchers, 2019). The study suggests that organisations are not using influencers' expertise as widely as the agencies and theory suggest they should. Table 5 presents the influencer activity in communication strategy and shows its use in the organisation (Enke & Borchers, 2019).

Table 5 Influencer roles in strategic communication

	Activity of the influencer	Description
org1	Content creator	Content created for an influencer's channels and for an organisation to use in other channels
	Multiplicator	An influencer shares at collaboration on his or her channel
	Protagonist	A meet-and-greet event for loyal customers
org2	Content creator	Content creation for influencer channels
	Multiplicator	An influencer shares the collaboration on his or her channel
	Protagonist	Modelling for a photoshoot
org3	Content creator	Content created for an influencer's channels and for an organisation to use in other channels
	Multiplicator	An influencer shares the collaboration on his or her channel
org4	Content creator	Content creation for influencer channels
	Multiplicator	An influencer shares at collaboration on his or her channel

As Enke and Borchers (2019) explained, the influencer, as a content creator, can create content (such as text, video and pictures) for himself or herself or for the organisation's use. The study shows this role as the most-used influencer role, though it is mostly used for the influencer's channels. When influencers act as multiplicators, they share the organisational message on their own channels and, as Table 5 shows, the influencers only shared the content of the collaboration in order to increase the visibility of the organisational message. However, to maintain the authenticity and trustworthiness of the collaborations, influencers should be careful not to compromise their style in this activity. A moderator could take part in online interaction that is relevant to the organisation, engaging on behalf of the organisation, but none of the studied organisations had used an influencer as a moderator. Lastly, influencers were only used as protagonists in a couple of cases (one acting as a meet-and-greet event host and one as a model for the organisation). A protagonist could be the main character for some content in the organisation, an event host (internally or externally), a workshop expert or a spokesperson.

Here, the unstructured planning was visible; Sundermann and Raabe (2019) argued the this was the case with influencer collaboration in many organisations. The organisations carefully chose the right fit with the influencer but then failed to consider how the influencer could be used during the collaboration to get the most out of the relationship.

The second research question was, 'What roles does an influencer fulfil in an organisation's strategic communication?' The theory suggested that the possible roles for the influencer were content creator, multiplicator, moderator and protagonist. The study found that the content creator role was used (but not to its full capacity) to create content for the organisation's use and that the protagonist role was used to some extent.

Furthermore, the influencer's many abilities suggest that several roles that could be fulfilled during the collaboration, so it would be important for an organisation to consider the purposes for which the influencer is to be used. One such role is that of the content creator, working individually or with the organisation; the content could be for the influencer's channels or for the organisation's use. Acting as a multiplicator, the influencer would share the organisational message in the content, ensuring the best possible outcome on the platform on which it appears. The influencer can also be used to take part in organisational events and workshops—online or offline, internal or external—as a protagonist, acting as an expert, a performer or a documentarian. Finally, the influencer can take part in relevant interaction as a moderator for the organisation in both online and offline channels.

Depending on the intended diversity of the relationship with the influencer, these roles can be used independently of each other or the influencer can combine multiple roles in a collaboration. A mix of these roles can lead to more effective outcomes than a single role would yield and thus should be used in the collaborations.

To summarise, the study gave insight into ways of improving actions ensuring brand fit with the influencer and the effective use of the roles that the influencer can fulfil. The organisations followed a process to ensure brand fit with an influencer, and the theory is actualised in parts of the process. It could be predicted that when the field is more settled, more connections to the theory will be found as the processes in organisations develop. In addition, this prediction could apply to the roles of the influencers as organisations find more ways to use them. Nevertheless, the findings in this study have connections to the theoretical models presented here; influencer marketing is settling into strategic communication.

5.1 Limitations and Future Research

The goal of this chapter was to gain an understanding of the influencer selection process and the ways in which influencers are used in organisations' communication strategies. This goal was achieved by thematic interviews and analysis. Thus, the study provides insights into the processes that organisations use when selecting influencers, the goals for different relationship lengths and the roles given to influencers during the collaborations.

However, the research findings have certain limitations that need to be considered in the interpretation. The findings are based on the interviews of certain organisations in the retail field in Finland and, therefore, cannot be generalised to companies in every field. In addition, the influencer concept was new to all the organisations, so the findings could be limited in that the organisations had not fully integrated influencer marketing into their strategies. Although the selection of the interviewees has limited the generalisation of the results, this study has gathered valuable information for future research. Studying other organisations from different fields and with different experience levels in regard to influencer marketing should be done in the future. Also, research on organisations that have been working with influencer in long-term collaborations should be implemented to understand the importance of the role in an organisation and how the role changes over time.

Appendix: Email to the Interviewees

The Themes and Preliminary Questions for the Interview

The interview is informal and focuses on gathering information broadly on the influencer collaborations in organisations. I have created preliminary questions, which might help you when preparing for the interview. The questions might not be asked in the same form or order in the interview.

Collaborations Between Influencers and the Organisation

Tell freely about previous collaborations.

- For example:
  - What processes did you use to select the influencers?

- How many collaborations has the organisation undertaken?
- Which channels did each collaboration use?
- What was the goal of each collaboration?
- What things did you consider before, during and after each collaboration?
- Where did you find the influencers?
- Please add anything else that comes to mind.

The Relationship Between Influencer and Organisation

- How is the fit of the influencer decided?

  - Values, reach, audience, other?
  - Do you consider the brand of the influencer?

- Are the collaborations long-or short-term?

  - Does each collaboration have its own reasons?

- What roles, jobs or tasks does the influencer fulfil during a collaboration?

Organisational Communication Strategy

- Do you plan the influencer collaborations as part of the communication strategy?

  - Per campaign, annually, for product launches, other?
  - Are they also visible in channels such as Google or Facebook?
  - Do you build a persona for the influencer?
  - Do you use an influencer agency?

- Have the influencer collaborations affected internal roles in the organisation?
- Describe a successful collaboration:

  - In your opinion, what important points made the collaboration successful?
  - What metrics do you use for collaborations?

References

Adams, J., Khan, H. T. A., & Raeside, R. (2014). *Research methods for business and social science students* (2nd ed.). SAGE Response.

Bakker, D. (2018). Conceptualising influencer marketing. *Journal of Emerging Trends in Marketing and Management, 1*(1), 79–87.

Borchers, N. S. (2019). Social media influencers in strategic communication. *International Journal of Strategic Communication, 13*(4), 255–260. https://doi.org/10.1080/1553118X.2019.1634075

Caudle, S. L. (2004). Qualitative data analysis. *Handbook of Practical Program Evaluation, 2*(1), 417–438.

Colliander, J., & Erlandsson, S. (2015). The blog and the bountiful: Exploring the effects of disguised product placement on blogs that are revealed by a third party. *Journal of Marketing Communications, 21*(2), 110–124. https://doi.org/10.1080/13527266.2012.730543

Deges, F. (2018). *Quick guide influencer marketing: Wie Sie durch Multiplikatoren mehr Reichweite und Umsatz erzielen.* Springer.

Djafarova, E., & Rushworth, C. (2016). Exploring the credibility of online celebrities' Instagram profiles in influencing the purchase decisions of young female users. *Computers in Human Behavior, 68*, 1–7. https://doi.org/10.1016/j.chb.2016.11.009

Enke, N., & Borchers, N. (2019). Social media influencers in strategic communication: A conceptual framework for strategic social media influencer communication. *International Journal of Strategic Communication, 13*(4), 261–277. https://doi.org/10.1080/1553118X.2019.1620234

Evans, N. J., Phua, J., Lim, J., & Jun, H. (2017). Disclosing Instagram influencer advertising: The effects of disclosure language on advertising recognition, attitudes, and behavioral intent. *Journal of Interactive Advertising, 17*(2), 138–149. https://doi.org/10.1080/15252019.2017.1366885

Freberg, K., Graham, K., McGaughey, K., & Freberg, L. A. (2011). Who are the social media influencers? A study of public perceptions of personality. *Public Relations Review, 37*, 90–92. https://doi.org/10.1016/j.pubrev.2010.11.001

Hou, M. (2018). Social media celebrity and the institutionalization of YouTube. *Convergence: The International Journal of Research into New Media Technologies, 25*(3), 534–553. https://doi.org/10.1177/1354856517750368

Johnstone, L., & Lindh, C. (2018). The sustainability-age dilemma: A theory of (un)planned behaviour via influencers. *Journal of Consumer Behaviour, 17*(1), e127–e139. https://doi.org/10.1002/cb.1693

Kapitan, S., & Silvera, D. (2016). From digital media influencers to celebrity endorsers: Attributions drive endorser effectiveness. *Marketing Letters, 27*(3), 553–567. https://doi.org/10.1007/s11002-015-9363-0

Keel, A., & Nataraajan, R. (2012). Celebrity endorsements and beyond: New avenues for celebrity branding. *Psychology & Marketing, 29*(9), 690–703. https://doi.org/10.1002/mar.20555

Macnamara, J. (2018). A review of new evaluation models for strategic communication: Progress and gaps. *International Journal of Strategic Communication, 12*(2), 180–195. https://doi.org/10.1080/1553118X.2018.1428978

Malhotra, N. K., Birks, D. F., & Wills, P. (2012). *Marketing research: An applied approach* (4th ed.). Pearson.

Ohanian, R. (1990). Construction and validation of a scale to measure celebrity endorsers' perceived expertise, trustworthiness, and attractiveness. *Journal of Advertising, 19*(3), 39–52. https://doi.org/10.1080/00913367.1990.10673191

Pöyry, E., Pelkonen, M., Naumanen, E., & Laaksonen, S. (2019). A call for authenticity: Audience responses to social media influencer endorsements in strategic communication. *International Journal of Strategic Communication, 13*(4), 336–351. https://doi.org/10.1080/1553118X.2019.1609965

Statista. (2022). Influencer marketing market size worldwide from 2016 to 2022. https://www.statista.com/statistics/1092819/global-influencer-market-size/. Accessed October 17, 2022

Sundermann, G., & Raabe, T. (2019). Strategic communication through social media influencers: Current state of research and desiderata. *International Journal of Strategic Communication, 13*(4), 278–300. https://doi.org/10.1080/1553118X.2019.1618306

Till, B. D., & Busler, M. (1998). Matching products with endorsers: Attractiveness versus expertise. *Journal of Consumer Marketing, 15*(6), 576–586. https://doi.org/10.1108/07363769810241445

Till, B. D., & Busler, M. (2000). The match-up hypothesis: Physical attractiveness, expertise, and the role of fit on brand attitude, purchase intent and brand beliefs. *Journal of Advertising, 29*(3), 1–13. https://doi.org/10.1080/00913367.2000.10673613

Personalization of Products and Sustainable Production and Consumption in the Context of Industry 5.0

Sebastian Saniuk ⓘ**, Sandra Grabowska** ⓘ**, and Mochammad Fahlevi** ⓘ

Abstract An important issue being developed in the concept of Industry 5.0 is sustainability as the most important direction of the modern world. We are already feeling the effects of environmental pollution, global warming and rising prices of energy resources. The societies of many European countries are beginning to realize the threat posed by the robbery of natural resources and excessive consumption. In general, the concept of Industry 5.0 should make modern industry more sustainable and human-centered. Hence, the chapter presents an analysis of consumer preferences for purchasing personalized production offered by implementing the concept of Industry 5.0 and ensuring sustainable consumption and production (SCP). The considerations were based on available literature sources and the results of the authors' own research conducted on a selected group of consumers, focusing on learning about expectations, consumer preferences for personalized products and conscious consumption. The most important achievement is the demonstration of a high level of consumer satisfaction with the purchase of personalized products and the positive impact of personalized production on sustainable consumption. At the same time, the importance of the development of the Industry 5.0 concept for supporting consumer behavior oriented toward sustainable consumption was emphasized.

Keywords Industry 5.0 · Sustainable consumption sustainable production · Conscious consumer · Personalization

S. Saniuk
Department of Engineering Management and Logistic Systems, University of Zielona Góra, Zielona Góra, Poland

S. Grabowska (✉)
Department of Production Engineering, Silesian University of Technology, Gliwice, Poland
e-mail: sandra.grabowska@polsl.pl

M. Fahlevi
Bina Nusantara University, Jakarta, Indonesia

1   Towards Industry 5.0 and Sustainable Development

The concept of Industry 4.0, which has been in development since 2011, is responsible for creating policies for building cyber-physical production systems to integrate information and operational technologies in enterprises and supply chains. The digital technologies used to build Industry 4.0 initially began to impose human strategies for dehumanizing industry and replacing humans with robots and intelligent, autonomous machines and devices. The first mention of dehumanization problems in Industry 4.0 appeared in publications by Romero et al. (2016a, 2016b). At that time, it was noted that there was a need for symbiosis between humans and new technologies, that there was a need to use the human mind to cooperate with intelligent machines and use its potential in the production process. The authors proposed introducing the human factor into cyber-physical systems. Newly designed systems should be humanized and designed as Human Cyber-Physical System (H CPS). This is how the concept of Industry 5.0 emerged, dictated primarily by the need to reveal the role of humans in cyber-physical systems as supervisors and decision-makers.

A positive effect of the Fourth Industrial Revolution and the widespread digitization of production systems is the move away from the need to produce products on a large scale simply because of low unit cost. Changing the structure of the workforce and redeploying workers to other spheres of the economy with the high productivity levels of such systems and the high level of flexibility makes it possible to change production strategies. Currently, it is possible to produce fewer products at the same time more tailored to customer needs without compromising the company's bottom line (Demir et al., 2019; Özdemir & Hekim, 2018; Xu et al., 2021).

In the development of the fourth industrial revolution, in addition to emphasizing the human factor, the very important need to take into account sustainability, accountability and safety has been noted (Longo et al., 2020). The concept of Industry 5.0 first appeared in 2020 documents among the participants of a conference organized by one of the subcommittees on Research and Innovation of the EC Commission. In general, the Industry 5.0 concept should make modern industry more sustainable and human-centered (Nahavandi, 2019). The idea of Industry 5.0 mainly focuses on the interaction between humans and machines and the creation of healthy relationships between them (Özdemir & Hekim, 2018). Humans should work in symbiosis with machines and should be connected to smart factories through smart devices (Huang et al., 2022). The world of technology, mass personalization and advanced manufacturing is undergoing a rapid transformation. Smart machines and devices, thanks to the development of artificial intelligence, should be connected to the human mind via a brain-machine interface (Leng et al., 2022). Today, robots are intertwined with the human brain and work as a collaborator, not a competitor (Nahavandi, 2019). At the same time, they should constantly learn ethical behavior from humans and, above all, not endanger humans through autonomous behavior.

An important issue being developed in the concept of Industry 5.0 is sustainability as the most important direction of the modern world. We are already feeling the effects of environmental pollution, global warming and rising prices of energy resources. The societies of many European countries are beginning to realize the threat posed by the plundering of natural resources and excessive consumption. The modern challenge, therefore, is to reconcile economic growth and the maintenance of quality of life with concern for the environment. One of the significant problems of modern economies and societies is the reduction of energy consumption derived from coal and gas and the reduction of overall consumption of goods and services, which is reflected in new patterns of social behavior. The level of modern consumption contributes to serious environmental problems manifested in climate change (global warming), degradation of the global ecosystem, resource depletion, biodiversity depletion or water, air and soil pollution, but also causes social stratification (Akundi et al., 2022; Javaid & Haleem, 2020).

The challenge for today's companies is therefore to continuously improve products, optimizing production technologies to produce more products with longer product life cycles and more customer-focused products with the least possible raw material consumption and environmental impact. To achieve this goal, national governments are implementing a sustainable consumption and production (SPC) strategy. The main goal of this strategy is to reduce consumerism, which is particularly responsible for the excessive consumption of natural resources. It is necessary to take advantage of the opportunities offered by the Industry 5.0 concept and formulate new marketing strategies in such a way as to enable sustainability and minimize the effects of excessive consumption (Fraga-Lamas et al., 2021; Leng et al., 2022; Saniuk et al., 2022).

The solution is to introduce new patterns of quality of life and the idea of well-being, especially in developed countries, consisting of, among other things, share economy, circular economy or personalization of production (products with an extended life cycle). Hence, there is increasing talk of so-called sustainable consumption patterns, which is a form of consumption directly related to the concept of sustainable development, oriented towards long-term socio-economic goals, especially in terms of positive environmental impact (Promoting Sustainable Consumption 2008) (Ghobakhloo et al., 2022). Hence, the chapter presents an analysis of consumer preferences for purchasing personalized production offered through the implementation of the Industry 5.0 concept and the provision of sustainable consumption and production (SCP). The considerations were based on available literature sources and the results of the authors' own research conducted on a selected group of consumers, focusing on learning about expectations, consumer preferences for personalized products and conscious consumption. The most important achievement is the demonstration of a high level of consumer satisfaction with the purchase of personalized products and the positive impact of personalized production on sustainable consumption. At the same time, the importance of the development of the Industry 5.0 concept for supporting consumer behavior oriented toward sustainable consumption was emphasized (Leng et al., 2022).

2   Sustainable Production and Consumption

The steadily worsening state of environmental pollution, excessive consumption of natural resources and the noticeable effects of global warming have resulted in the development strategy for the World to 2030. In 2015, all 193 UN member states unanimously adopted the document "Transforming our world: the 2030 Agenda for Sustainable Development" containing 17 Sustainable Development Goals and associated 169 specific tasks to be achieved by 2030. Indicators have been established for each task, so that progress toward the goals is monitored worldwide. Particularly noteworthy, by virtue of the chapter's considerations, is Goal 12, "Responsible Consumption and Production," which is intended to ensure patterns of sustainable consumption and production for the future. Successful implementation of this goal requires a systematic approach and cooperation among actors along the value chain, from producers to consumers. The process also includes consumer education activities to raise awareness of sustainable consumption and change their lifestyles. Table 1 shows the tasks under the "responsible consumption and production" objective and relates them to the area that these tasks will have the greatest impact on in terms of sustainable production and consumption. The tasks should mostly relate to the Economy 4.0 as a whole. Some of them should be implemented in terms of Industry 4.0 and Society 5.0.

Sustainable consumption is based on three principles (Jackson, 2014):

- Economic rationality—economic optimization in the choice of goods,
- Ecological rationality—the selection of goods that do the least harm to the environment,
- Social rationality—the choice of goods that solve social problems or do not exacerbate them.

The main objective of the EU environmental policy on sustainable production and consumption is to create consumer and producer behavior aimed at increasing the environmental performance of products throughout their life cycle, creating demand for environmentally friendly products and manufacturing technologies and, equally importantly, for consumers to make informed purchases (European Parliament). An important premise of this policy is to achieve a compromise between increasing the quality of life of society and respect for environmental aspects, minimizing the consumption of raw materials, and reducing waste and pollution (Kamani et al., 2019; Kravanja et al., 2015).

Unfortunately, the past rush to increase the quality of life and general prosperity has significantly contributed to environmental degradation and a significant reduction in natural resources. Excessive production and consumption has led to many problems in modern society. Excessive consumption is becoming an end in itself in modern society. The result is often excessive consumer indebtedness, lack of free time devoted to additional work and increasing stress levels caused by the desire for

Table 1 Tasks leading to sustainable consumption and production

No.	Tasks	
12.1	Implementing a 10-year framework of programs on sustainable consumption and production, all countries are taking action, with developed countries taking the lead, taking into account the development and capabilities of developing countries	Economy 4.0
12.2	By 2030, achieve sustainable management and efficient use of natural resources	Economy 4.0
12.3	By 2030, halve global per capita food losses at the retail and consumer level and reduce food losses in production and supply chains, including post-harvest losses	Industry 4.0 Society 5.0
12.4	By 2020, achieve environmentally sound management of chemicals and all wastes throughout their life cycle, in accordance with established international frameworks, and significantly reduce their release into the air, water and soil to minimize their adverse effects on human health and the environment	Industry 4.0
12.5	By 2030, significantly reduce waste generation through prevention, reduction, recycling and reuse	Industry 4.0
12.6	Encourage companies, especially large and transnational ones, to adopt sustainable practices and integrate sustainability information into their reporting cycle	Industry 4.0
12.7	Promote procurement practices that are sustainable, in line with national policies and priorities	Industry 4.0
12.8	By 2030, provide people around the world with adequate information and awareness of sustainable development and lifestyles in harmony with nature	Society 5.0
12.a	Supporting developing countries to strengthen their scientific and technological capacity to transition to more sustainable consumption and production patterns	Economy 4.0
12.b	Develop and implement tools to monitor the impact of sustainable development on sustainable tourism that creates jobs and promotes local culture and products	Industry 4.0
12.c	Rationalize inefficient fossil fuel subsidies that encourage wasteful consumption by eliminating market distortions in accordance with national conditions, including by restructuring taxation and phasing out these harmful subsidies, if any, to reflect their environmental impact, taking full account of the special needs and conditions of developing countries and minimizing any adverse impact on their development in a way that protects the poor and affected communities	Economy 4.0

Source cf. Agenda for Sustainable Development 2030

consumer goods that cannot be afforded. In addition, overconsumption inevitably leads to waste. Purchased goods are not fully utilized and their production is always associated with the consumption of a significant amount of energy and resources, resulting in unnecessary environmental pollution (Cho et al., 2018; Venkatesan et al., 2021).

Also of great importance in the production of these goods is the so-called water footprint, especially in the case of animal products. Wastefulness means that the purchase of new products is not dictated by a real need but only by the desire to have a newer model of a product, such as a phone, car or TV. Such behavior does not justify the unnecessary use of raw materials at the production stage. Especially if the previously used product is operational and fulfills its functions. A common term for the above phenomenon is consumerism, understood as a negative social attitude, which is characterized by the unjustified acquisition of material goods and services, the production of which contributes to the waste of natural resources (Lira et al., 2022).

Each of the previous industrial revolutions has been characterized by a different approach to the role of consumption in economic development. What has been important up to now has been the growth of production and thus the growth of consumption of goods and services. In the first industrial revolution, an important achievement was the introduction of mass production thanks to which the availability of more goods raised the standard of living of society. The second industrial revolution contributed to a greater variety of manufactured products, electrification contributed to the development of products powered by electricity. There was a further increase in the propensity to consume. The third industrial revolution has seen the rise of consumerism. Thanks to automation and robotization, there is a rapid increase in the flexibility and productivity of production systems. Mass production is being replaced by mass production. There is an increase in new product launches thanks to computer-aided design (CAD) systems and a shortening of the life cycles of individual products. Companies see in greater production an opportunity to grow and retain their employees. They are competing with new products that feature new, modern shapes or new functions. The development of marketing strategies aimed at acquiring new products and the frequent replacement of existing products has negatively affected environmental pollution and the consumption of natural resources. Only the fourth industrial revolution, thanks to the possibility of using intelligent machines and equipment and the possibility of high production flexibility, allows a high level of castomization, which is positively received by modern society, which today expects personalized products and is convinced of the negative effects of excessive consumption (Koc & Teker, 2019; Nuvolari, 2019; Popkova et al., 2019).

An additional motivating factor for implementing the idea of sustainable production and consumption is the effects of excessive production on the environment. Overconsumption of goods and services leads to excessive production of waste, including plastics or electro-waste that are difficult to manage. The concept of sustainable consumption and production is therefore an interpretation of the concept of sustainability, related to consumption (Tseng et al., 2020). Tseng et al. (2018) adds that a responsible consumer is one who consciously takes this paradigm into account in the process of consumer decision-making. Thus, an informed consumer is an equivalent business partner. Sustainable consumer behavior along the value chain is shown in Table 2.

Today, the challenge for businesses and Economy 4.0 is to reconcile the fastest possible economic growth and the drive to improve the quality of life of society

Table 2 Sustainable consumer behavior at each stage of the value chain

A link in the value chain	Consumer behavior
Design	Co-design (e.g., through user-driven innovation or crowdsourcing)
Manufacturing	Prosumption (producing or co-producing goods and services)
Distribution and sales	Choosing a sustainable delivery method (online shopping)
Consumption and use	• Limiting oneself (giving up some needs or products, paying attention to technical rather than moral consumption of products) • Eliminating waste • Buying environmentally friendly products (made of natural raw materials, certified and with eco-labels) • Using reusable products • Buying virtual products (books, music, movies) • Buying remanufactured products • Buying services instead of products • Sharing • Donating (giving "second life to products") • Lending • Repairing • Buying from manufacturers
Recycling and recovery	• Use in new applications • Upcycling and downcycling • Collection and segregation of waste

Source Adapted from Wilk (2015)

with environmental protection. To achieve this goal, it is necessary to balance the three aspects of development: economic, social and environmental. This necessitates the continuous improvement of products, optimization of production technology, so that with the least possible consumption of raw materials and impact on the environment, to produce products with the best possible performance, economically viable products. The answer to the problems caused by waste of resources is also personalized production leading to sustainable consumption. This approach requires customer involvement at every stage of the product life cycle (Bag & Pretorius, 2020; Dantas et al., 2021; Kaz et al., 2019).

3 Personalization as a Result of Modern Customer Expectations

Companies operating in today's market are beginning to understand the need for change, not only in the area of the aforementioned assumptions of the idea of Industry 4.0, the implementation of modern digital technologies and intelligent cyber-physical systems, but also sustainable development including alignment with sustainable production and consumption strategies. There is also a need for a completely new, more modern and innovative approach to production and business management, which will radically increase customer orientation with the support of digital smart technologies.

Conscious customers today expect products better tailored to their personal preferences, tastes, needs and lifestyles. Easy access to social networks, the creation of socially responsible behavior and the recently fashionable trend of conscious consumption are changing the expectations of today's customers. Customers increasingly want to have more influence on the product they order by being able to tailor it to their own needs. The automotive industry is a prime example of this. About 80% of new cars ordered are subject to configuration by the customer at the order stage. Functionality and price are important factors in selecting the options chosen. This forces competing companies to increase the productivity and flexibility of their production systems and to orient themselves to a higher level of building interaction between the company and the customer. In the future, interaction building is aimed at enabling the customer to co-design a new product through the use of intuitive computer-aided design systems that simultaneously generate control codes dedicated to a given cyber-physical system of intelligent machines responsible for manufacturing and delivering the product to the customer (Govindan, 2018).

Chen et al. (2019) defines personalization as any customization of a product (its features, method of distribution and even promotion) to meet individual customer needs. Businesses are therefore required to respond quickly to customer needs, in terms of developing a personalized product, delivering the order in a timely manner and ensuring a low purchase price. One of the modern forms of enterprise communication with customers is customization. Mass customization involves personalization of product offerings and services on a large scale, which is made possible by the rapid development of automation, robotization and digitization of production and logistics processes, as well as in-depth knowledge of consumer needs and preferences. Its goal is to optimally meet consumer needs through better interaction during the process of designing new products (Suzić et al., 2018). The main goal of castomization is to produce customized products with production costs and price levels close to those of mass-produced products (Pallant et al., 2020).

Lampel and Mintzberg, depending on the degree of customer involvement in the new product development process, distinguish between five mass customization strategies, which are pure customization, tailored customization, segmented standardization, and pure standardization (Lampel & Mintzberg, 1996). Pure customization is a strategy in which there is a strong relationship between the manufacturer and the customer. The customer is involved in all stages of production, including the design stage of a new product. He has a very strong influence on the final shape and functionality of the product. In this strategy, products are completely individualized, virtually unique. The second strategy, so-called tailored customization, means that the customer has an influence on the selection of dimensions, shape of standard product components. The product is tailored to the individual preferences of the buyer to a limited extent depending on the possible production capacity of the manufacturer. The third strategy, in which the customer is only involved in the assembly or distribution phase of the product, is standardized castomization. The product is modified and configured according to a list of standard options. This most often involves the choice of color, equipment options or materials used. The last type of castomization strategy is pure standardization, where the customer has no influence on the product (Lampel & Mintzberg, 1996; Mintzberg et al., 2003).

Currently, the acquisition of competitive advantages by companies goes beyond the classical norms of competing on price, availability and quality. The competitiveness of companies that based their strategy solely on the area of production and logistics is now declining. Aspects of sustainable production and corporate social responsibility are slowly gaining importance. Growing consumer awareness and today's fashionable trends of caring for the environment, numerous promotional campaigns warning against excessive environmental pollution are changing the modern consumer's approach to product offerings. We are increasingly interested in the manufacturer's impact on sustainable development, we are interested in the manufacturer's approach to environmental and social aspects. Those companies that are socially responsible are gaining (Gareche et al., 2019).

The research presented in Saniuk et al. (2020) presents the results of a consumer survey, which shows that the vast majority of consumers surveyed already pay attention to whether the manufacturer is socially responsible when making purchasing decisions. Only for 3% of respondents this aspect does not matter at all (see Fig. 1). The results obtained testify to the ever-increasing high level of consumer awareness of sustainability. Activities related to the creation of correct behavior of conscious purchase of environmentally safe products manufactured by producers with a high level of Corporate Social Responsibility (CSR) are beginning to bear fruit.

The main objective of the survey conducted by the authors was also to demonstrate the interest of today's customers in personalized production. According to the authors, personalization can help reduce overconsumption and influence the life cycle of a better-fitting product. The survey was conducted on a group of 504 potential customers and consumers representing Polish society. Assuming a confidence level of 0.99 and an error of 10%, it was determined that the minimum size of the general population should be 166 customers. The survey was conducted using the Computer Assisted Web Interview (CAWI) method in 2019. Most respondents were

Fig. 1 The importance of corporate social responsibility of the producer for today's customers. *Source* Own elaboration

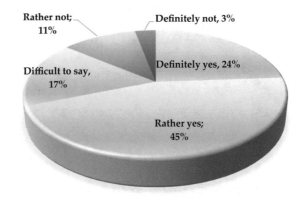

from large and medium-sized cities (59%). It is worth noting that the majority of respondents rated their material situation as good (64.1%) and sufficient (23.24%). Very good material situation was declared by about 13.9% of respondents. 52% of respondents were interested in buying personalized products. Most often such products are purchased by consumers aged 19–25 (38.04%), 26–35 (14.95%), 36–45 (15.76%), rarely by consumers aged 56–67 (7.61%), very rarely such purchases are made by consumers over 67 (1.36%). Based on these results, it can be seen that the most active group of consumers purchasing personalized products are young people aged 19–45. This is understandable given the greater propensity of young people to succumb to today's fashionable trends especially presented on social media and promoted by well-known influencers. In addition, younger generations are better versed in the use of the e-commerce environment and digital e-service technologies they are eager to use.

Observations of today's consumers show that electronics and consumer electronics are most often personalized (42% of respondents), 39% of respondents indicated that they personalize custom dishes, and various accessories (calendars, cases, etc. (33% of respondents). Consumers also like to personalize clothing and footwear, jewelry and children's toys. They are less likely to use personalization when buying cosmetics (17%), cars (15.3%), household appliances and software (12.8%) (Saniuk et al., 2020). Figure 2 shows the types of personalized products that are most often indicated by today's consumers.

Also noteworthy is the expressed expectation of potential customers regarding the level of involvement in the production process of a personalized product. Respondents

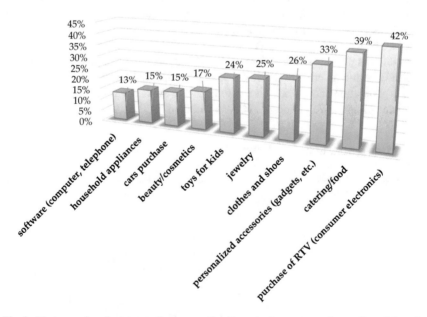

Fig. 2 The types of products most often personalized by today's customers. *Source* Own elaboration

clearly condition their interest in their participation in the process of creating new products on the types of products. Pure customization is the most expected strategy for the purchase of food services (38% of respondents), jewelry (42%), furniture and garden equipment (37%), clothing and shoes (44%). Tailored customization was most often chosen by customers buying: furniture and garden equipment (37%), food services (31%), clothes and shoes (31%), car accessories (32%). Standardized castomization was indicated most often by respondents when buying: electronics (36%), toys (32%), cosmetics (31%), car accessories (30%), common products (30%). In contrast, pure standardization applies only to purchases of common use products (35%). Responding customers were given the opportunity to read a description of the different castomization strategies for a better understanding of the differences between them (Grabowska & Saniuk, 2021; Saniuk et al., 2020). The detailed results of the survey in this area are shown in Fig. 3.

An interesting reason for customers to personalize products is their emotional involvement in the creation of a new product that will express their personal preferences and expectations. In the survey, more than 56% of respondents stressed that the main reason for purchasing personalized products is its uniqueness (uniqueness). In addition, their purchase decisions are also dictated by the higher comfort of the product (about 48% of respondents), the possibility of deciding on the final shape

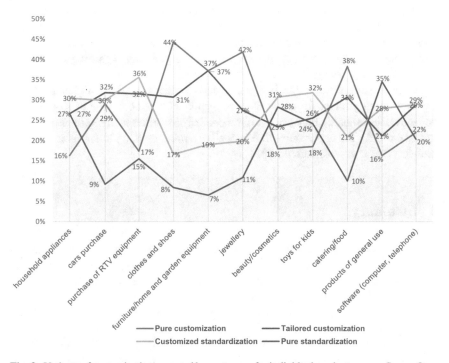

Fig. 3 Variants of customization expected by customers for individual product groups. *Source* Own elaboration

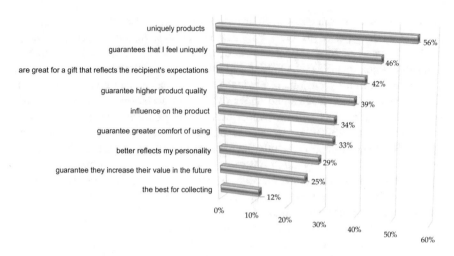

Fig. 4 Main reasons for choosing personalized products. *Source* Own elaboration

of the product (46% of respondents) and the possibility of better fitting the product acting as a gift (about 42% of respondents). Also noteworthy is the identification by as many as 39% of respondents of a guarantee of higher product quality with personalization. Personalized products better reflect the personality of the customer, such a statement is confirmed by about 34% of respondents (Grabowska & Saniuk, 2021). Figure 4 shows the detailed responses of respondents regarding the reasons for personalizing products.

An interesting observation is the stated willingness to pay a higher price for a personalized product for about 39% of respondents. About 58% of respondents depend on the amount of the price difference. Only 3% of respondents are not willing to pay more for the opportunity to personalize a product than for a similar standard product (Grabowska & Saniuk, 2021).

4 Summary

Today's customers are accustomed to a wide range of products and the privileged position they are in, they expect more than just the best quality product at the lowest price. Unfortunately, the recent noticeable shortage of raw materials and rising energy costs are forcing manufacturers to change their operating strategy. In the future, they should produce fewer products, try to design products with longer life cycles and apply circular economy principles. At the same time, customers who are aware of sustainable production and consumption look to the personalization of products to maintain an adequate level of quality of life with rational use of limited resources. Providing personalized products can guarantee benefits for both sides of the transaction. A customer who is satisfied with the product he or she receives will be more loyal, which in turn can translate into revenue stability for the manufacturer. For today's manufacturing companies this means, the need for changes in the way production is managed, an orientation towards personalized products and a higher level of business networking, which should build a direct relationship with the consumer involving him in the process of designing and manufacturing products preferably using ICT. In the long term, this could mean a shift, especially by small and medium-sized enterprises, to the servitization of industrial production.

Sustainable consumption and production combines, on the one hand, the need to meet needs, improve the quality of life, and on the other hand, improve resource efficiency, increase the use of renewable energy sources, minimize waste. Integration of these elements is the main goal of modern economies, which want to provide the same or better services to meet the basic needs of life and aspire to improve the quality of life while constantly reducing environmental damage and risks to human health. Modern businesses are forced to meet the high demands of consumers, who increasingly expect tailored products and are increasingly aware of the negative effects of over-consumption.

The consumer survey results presented indicate a high level of satisfaction with the purchase of personalized products and a strong interest in increasing consumer involvement in the product design and manufacturing process. Increased satisfaction translates into a level of satisfaction with the long-term use of personalized products and an overall reduction in consumption. Noteworthy is the increased awareness of the modern consumer and attention to the social responsibility of producers. Social media is creating new behavior based on conscious consumption, the development of sharing economy behavior and care for the environment. There is no doubt that such a profile of the modern consumer has been greatly influenced by the development of the Internet, mobile telephony, market globalization and many other determinants rooted in the macro environment.

The fourth industrial revolution has influenced a greater level of integration between customer and manufacturer resulting in the possibility of a high level of production personalization. Customer involvement in the product design and manufacturing process should be considered not only in terms of cost minimization and improved customer orientation, but also in terms of an opportunity to change consumer behavior oriented toward sustainable production and consumption. Today's companies will have to change their orientation from product to service. Offer a high level of design, manufacturing and logistics services offering personalized products manufactured in sophisticated enterprise networks using Industry 4.0 technologies. The Fourth Industrial Revolution thus means that cyber-physical systems are actively interacting with customers, and the level of product creation is shifted from the perspective of creating a physical product to creating new experiences and building customer satisfaction. When buying a personalized product, the customer makes that purchase with greater awareness and feels greater satisfaction with the purchase, which in the long run leads to a reduction in overall consumption levels and an increase in sustainable consumption.

The research conducted prompts further studies in the future to demonstrate the impact of personalized manufacturing on extending the life cycle of products, reducing overall consumption, and overall reduction of energy and natural resource consumption, especially in the context of the conscious purchasing decisions of today's consumers.

References

Agenda for Sustainable Development 2030. https://www.undp.org/sustainable-development-goals?utm_source=EN&utm_medium=GSR&utm_content=US_UNDP_PaidSearch_Brand_English&utm_campaign=CENTRAL&c_src=CENTRAL&c_src2=GSR&gclid=Cj0KCQjw1 66aBhDEARIsAMEyZh40HO8qoM3HEjfdcGynBNAf_iBblxuJiplqF1oCFhGXc_lXoehqZ5w aAutSEALw_wcB. Accessed October 14, 2022

Akundi, A., Euresti, D., Luna, S., Ankobiah, W., Lopes, A., & Edinbarough, I. (2022). State of Industry 5.0—Analysis and identification of current research trends. *Applied System Innovation, 5*(1), 27.

Bag, S., & Pretorius, J. H. C. (2020). Relationships between Industry 4.0, sustainable manufacturing and circular economy: Proposal of a research framework. *International Journal of Organizational Analysis.*

Chen, Q., Feng, Y., Liu, L., & Tian, X. (2019). Understanding consumers' reactance of online personalized advertising: A new scheme of rational choice from a perspective of negative effects. *International Journal of Information Management, 44*, 53–64.

Cho, Y. N., Soster, R. L., & Burton, S. (2018). Enhancing environmentally conscious consumption through standardized sustainability information. *Journal of Consumer Affairs, 52*(2), 393–414.

Dantas, T. E., De-Souza, E. D., Destro, I. R., Hammes, G., Rodriguez, C. M. T., & Soares, S. R. (2021). How the combination of Circular Economy and Industry 4.0 can contribute towards achieving the Sustainable Development Goals. *Sustainable Production and Consumption, 26*, 213–227.

Demir, K. A., Döven, G., & Sezen, B. (2019). Industry 5.0 and human-robot co-working. *Procedia Computer Science, 158*, 688–695.

Fraga-Lamas, P., Lopes, S. I., & Fernández-Caramés, T. M. (2021) Green IoT and edge AI as key technological enablers for a sustainable digital transition towards a smart circular economy: An Industry 5.0 use case. *Sensors, 21*(17), 5745.

Gareche, M., Hosseini, S. M., & Taheri, M. (2019). A comprehensive literature review in competitive advantages of businesses. *International Journal of Advanced Studies in Humanities and Social Science, 6*(4), 312–329.

Ghobakhloo, M., Iranmanesh, M., Mubarak, M. F., Mubarik, M., Rejeb, A., & Nilashi, M. (2022). Identifying Industry 5.0 contributions to sustainable development: A strategy roadmap for delivering sustainability values. *Sustainable Production and Consumption, 33*, 716–737.

Govindan, K. (2018). Sustainable consumption and production in the food supply chain: A conceptual framework. *International Journal of Production Economics, 195*, 419–431.

Grabowska, S., & Saniuk, S. (2021). Modern marketing for customized products under conditions of fourth industrial revolution. In J. Duda, & A. Gąsior (Eds.), *Industry 4.0: A glocal perspective* (pp. 1–13).

Huang, S., Wang, B., Li, X., Zheng, P., Mourtzis, D., & Wang, L. (2022). Industry 5.0 and Society 5.0—Comparison, complementation and co-evolution. *Journal of Manufacturing Systems, 64*, 424–428.

Jackson, T. (2014). *Sustainable consumption.* Edward Elgar Publishing.

Javaid, M., & Haleem, A. (2020). Critical components of Industry 5.0 towards a successful adoption in the field of manufacturing. *Journal of Industrial Integration and Management, 5*(03), 327–348.

Kamani, M. H., Eş, I., Lorenzo, J. M., Remize, F., Roselló-Soto, E., Barba, F. J., & Khaneghah, A. M. (2019). Advances in plant materials, food by-products, and algae conversion into biofuels: Use of environmentally friendly technologies. *Green Chemistry, 21*(12), 3213–3231.

Kaz, M., Ilina, T., & Medvedev, G. A. (2019). *Global economics and management: Transition to Economy 4.0.* Springer. https://doi.org/10.1007/978-3-030-26284-6

Koc, T. C., & Teker, S. (2019). Industrial revolutions and its effects on quality of life. *PressAcademia Procedia, 9*(1), 304–311.

Kravanja, Z., Varbanov, P. S., & Klemeš, J. J. (2015). Recent advances in green energy and product productions, environmentally friendly, healthier and safer technologies and processes, CO_2 capturing, storage and recycling, and sustainability assessment in decision-making. *Clean Technologies and Environmental Policy, 17*(5), 1119–1126.

Lampel, J., & Mintzberg, H. (1996). Customizing customization. *Sloan Management Review, 38*(1), 21–30.

Leng, J., Sha, W., Wang, B., Zheng, P., Zhuang, C., Liu, Q., & Wang, L. (2022). Industry 5.0: Prospect and retrospect. *Journal of Manufacturing Systems, 65*, 279–295.

Lira, J. S. D., Silva Júnior, O. G. D., Costa, C. S. R., & Araujo, M. A. V. (2022). Fashion conscious consumption and consumer perception: A study in the local productive arrangement of clothing of Pernambuco. *BBR. Brazilian Business Review, 19*, 96–115.

Longo, F., Padovano, A., & Umbrello, S. (2020). Value-oriented and ethical technology engineering in Industry 5.0: A human-centric perspective for the design of the factory of the future. *Applied Sciences, 10*(12), 4182.

Mintzberg, H., Ghoshal, S., Lampel, J., & Quinn, J. B. (2003). *The strategy process: Concepts, contexts, cases.* Pearson Education.

Nahavandi, S. (2019). Industry 5.0—A human-centric solution. *Sustainability, 11*(16), 4371.

Nuvolari, A. (2019). Understanding successive industrial revolutions: A "development block" approach. *Environmental Innovation and Societal Transitions, 32*, 33–44.

Özdemir, V., & Hekim, N. (2018). Birth of Industry 5.0: Making sense of big data with artificial intelligence, "the internet of things" and next-generation technology policy. *Omics: A Journal of Integrative Biology, 22*(1), 65–76.

Pallant, J., Sands, S., & Karpen, I. (2020). Product customization: A profile of consumer demand. *Journal of Retailing and Consumer Services, 54*, 102030.

Popkova, E. G., Ragulina, Y. V., Bogoviz, A. V. (2019). Fundamental differences of transition to Industry 4.0 from previous industrial revolutions. In *Industry 4.0: Industrial revolution of the 21st century* (pp. 21–29). Springer.

Romero, D., Bernus, P., Noran, O., Stahre, J., & Fast-Berglund, Å. (2016a, September). The operator 4.0: Human cyber-physical systems & adaptive automation towards human-automation symbiosis work systems. In *IFIP International Conference on Advances in Production Management Systems* (pp. 677–686). Springer.

Romero, D., Stahre, J., Wuest, T., Noran, O., Bernus, P., Fast-Berglund, Å., Gorecky, D. (2016b, October). Towards an operator 4.0 typology: A human-centric perspective on the fourth industrial revolution technologies. In *Proceedings of the International Conference on Computers and Industrial Engineering (CIE46)*, Tianjin, China (pp. 29–31).

Saniuk, S., Grabowska, S., & Gajdzik, B. (2020). Social expectations and market changes in the context of developing the Industry 4.0 concept. *Sustainability, 12*, 1–20. https://doi.org/10.3390/su12041362

Saniuk, S., Grabowska, S., & Straka, M. (2022). Identification of social and economic expectations: Contextual reasons for the transformation process of Industry 4.0 into the Industry 5.0 concept. *Sustainability, 14*(3), 1391.

Suzić, N., Forza, C., Trentin, A., & Anišić, Z. (2018). Implementation guidelines for mass customization: Current characteristics and suggestions for improvement. *Production Planning & Control, 29*(10), 856–871.

Tseng, M. L., Chiu, A. S., & Liang, D. (2018). Sustainable consumption and production in business decision-making models. *Resources, Conservation and Recycling, 128*, 118–121.

Tseng, M. L., Chiu, A. S., Liu, G., & Jantaralolica, T. (2020). Circular economy enables sustainable consumption and production in multi-level supply chain system. *Resources, Conservation and Recycling, 154*, 104601.

Venkatesan, M., Dreyfuss-Wells, F., Nair, A., Pedersen, A., & Prasad, V. (2021). Evaluating conscious consumption: A discussion of a survey development process. *Sustainability, 13*(6), 3339.

Wilk, I. (2015). Konsument zrównoważony jako segment odniesienia dla działań marketingowych przedsiębiorstwa. Zeszyty Naukowe Uniwersytetu Szczecińskiego. *Problemy Zarządzania, Finansów i Marketingu, 38*, 183–192.

Xu, X., Lu, Y., Vogel-Heuser, B., & Wang, L. (2021). Industry 4.0 and Industry 5.0—Inception, conception and perception. *Journal of Manufacturing Systems, 61*, 530–535.

Energy in the Era of Industry 5.0—Opportunities and Risks

Marius Gabriel Petrescu, Adrian Neacșa, Eugen Laudacescu, and Maria Tănase

Abstract Industry 5.0 requires the resettlement of man (the worker) at the center of industrial processes. The concept of Industry 5.0 requires the use of advanced technologies to support man in his actions, to help him to progress and to offer him the solutions for his needs and interests. Industry 5.0 embraces the idea of a sustainable industry and opens the way to a healthier future, the reference point being considered—and here comes the novelty in relation to Industry 4.0—"a totally sustainable society". In this context we cannot—and must not—avoid discussing the sustainability of the industrial sector and its conditioning on the sustainability of the planet's energy resources. The energy transition offers solutions to protect the environment, but it raises economic, social and technical issues that companies, public authorities, financial institutions and researchers need to address. The complexity of the energy transition phenomenon requires the involvement of companies, consumers, portfolio investors and the education system, which must encourage a change of mentality and improved behavior.

Keywords Industry 5.0 · Resources · Energy · Sustainability · Social

1 Introduction

Humanity, throughout its evolution, has constantly sought to increase the comfort of its existence, to develop products that respond to increasingly ambitious needs and to expand the exploitation of the resources offered by nature.

Just as the Earth is sustained by energy—energy it contains as well as energy provided by the Universe—the man and his activities rely on the consumption of energy from various sources.

M. G. Petrescu (✉)
Petroleum-Gas University of Ploiești and Romanian Agency for Quality Assurance in Higher Education (ARACIS), Bucharest, Romania
e-mail: pmarius@upg-ploiesti.ro

A. Neacșa · E. Laudacescu · M. Tănase
Petroleum-Gas University of Ploiești, Ploiești, Romania

Many times the man has violated the limit of natural balance by exploiting the resources excessively. Often, these excesses were blamed on the so-called industrial revolutions.

We are now facing an industrial revolution—Industry 5.0—which, as it were, is trying to correct the mistakes of the past, looking for solutions in the direction of balancing the relationship between the man and the nature, but also the relationships between people.

What Industry 5.0 brings new, what it tries to correct in the context of the depletion of energy resources, is briefly presented—as the opinion of the authors—in this work.

2 Industry 5.0—Innovative Concept or a Complement to the Industry 4.0 Concept?

Industry 4.0 emerged, definitively, as a consequence of consumers' growing interest in products.

Starting from manufacturing in multiple phases of processing raw materials, using resources specific to the industrial production (human, knowledge, procedures, energy, etc.), INDUSTRY 4.0 also calls for the extensive use of computing resources together with an extensive use of communications (mainly IoT—Internet of Things that integrates through connections "objects" such as sensors, machines, etc.) to achieve a high level of flexibility and adaptability of manufacturing (IMT Bucharest, 2017).

In a market economy, the demand is the determining factor for production volume, product typology and their sustainability. Consequently, starting from 2011 (the year in which, for the first time, in Germany, the concept of Smart Manufacturing was approached, which is considered to be the fourth, concept representing the current stage, the fourth stage of the industrial revolution—Industry 4.0), there is an emphasis on production and product efficiency, and are implemented solutions that, beyond ensuring the production growth, address the entire product life cycle, from the design phase to recycling. The concerns of research-innovation field are now widely applicable, with laboratory activity being interconnected with the actual production. To increase the productivity, the assortment diversification and the implementation in products of some facilities requested and aimed directly at the general public, it appealed to:

- the implementation of predictive/proactive concepts/strategies, which ensure the direct transfer of laboratory results to the manufacturing;
- flexibility, robotization and optimization of manufacturing;
- optimization of product marketing;
- digitization and computerized assistance of the product throughout its entire life cycle.

We may consider, at some point, that such behavior—whether we are talking about production or consumption—is extravagant and exhausting. When we say this we are not referring to the evidence of current human societies, namely: the deepening of social differentiation; depletion of the (limited) resources available to the planet; the ecological consequences of excessive consumption.

Beyond the beneficial effects of applying the Industry 4.0 concept, consisting of:

- stimulation of innovative/inventive thinking;
- stimulation of entrepreneurship;
- the transfer of the results of leading industries in everyday life,

we must also recognize the previously mentioned negative impact consequences, the causes of which we can consider to be, mainly:

- the lack of coherent strategies—at the regional and global level—regarding the natural balance;
- the gap between economic and social policies;
- the struggle of the nations of the world to conquer the markets.

Of course, the created situations should not be judged simplistically, in the sense that the economic environment does not have the capacity to responsibly manage the challenges of this industrial revolution—Industry 4.0. The political factor—at the level of each nation and at the world level—is decisive for the correct promotion and for keeping economic and industrial phenomena under control and stopping any form of excess.

During its ten years of life, Industry 4.0 focused less on those principles that referred to social equity and economic sustainability and favored the promotion of measures regarding digitization and computerized manufacturing assistance, increasing the production efficiency and flexibility (Fig. 1). The concept of Industry 5.0 comes with corrections to the policies promoted by Industry 4.0, bringing to the fore the importance of research and innovation in support of a long-term, sustainable industry that serves humanity and conserves the resources offered by the planet (European Commission, 2021).

Historically speaking, it can be said that, starting from 2019, the global crisis generated by the Covid 19 pandemic was superimposed on top of the ecological crisis—presented simplistically in the media and in propaganda documents through the alarming increase in the volume of greenhouse gases in the atmosphere. Crossing over the initial moment of this crisis, characterized by hesitation and incoherence specific to a new threat to humanity, the innovative capacity, the efficiency of laboratory research, the flexibility of production and health systems—all of which, it must be said, are consequences and positive features of Industry 4.0—made it possible to find solutions—for the moment or long-term—for the considerable reduction of this scourge.

The period February–March 2021 offered to the humanity a new challenge as a result of the start of the war in Ukraine. The increase in intensity of the consequences of the war is evident, from one day to the next. The blockage of markets and transactions affects producers and consumers alike.

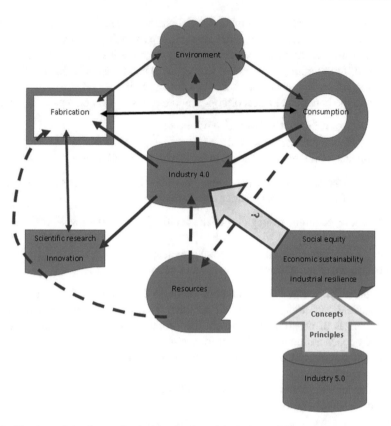

Fig. 1 The determining factors for the introduction of the Industry 5.0 concept (authors' view)

The crisis of energy and raw materials—as a result of excessive and, why not, irresponsible consumption—is accentuated and brings to the fore the issue of economic and social sustainability.

Through this review of the events of the last years—keeping us on the background of the fourth industrial revolution—we tried to highlight the advantages and the disadvantageous consequences that Industry 4.0 offered us (Fig. 2). Through this analysis, the authors try to motivate—if it is still the case—the transition to a new industrial revolution—Industry 5.0.

The concept of Industry 5.0 has developed in response to the social and environmental needs identified since 2020. In the sense of Industry 5.0, the industry is considered a loyal and beneficial tool for humanity if it responds to the three major demands: social, environmental and societal. The essence of Industry 5.0 is based on the symbiosis between the three segments: technological, social and ecological (Grabowska et al., 2022). Industry 5.0 focuses on the three factors (Fig. 3): human-centered development, sustainability and resilient development (Grabowska et al., 2022; Felsberger & Reiner, 2020; Romero et al., 2016).

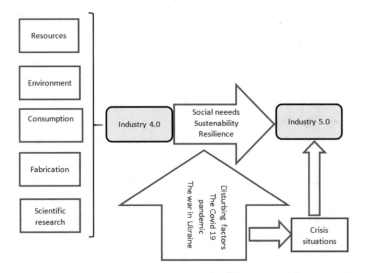

Fig. 2 The actual context of the promotion of Industry 5.0 (authors' view)

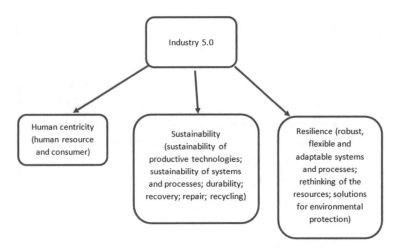

Fig. 3 The basic principles of Industry 5.0 (authors' view)

Industry 5.0 requires the resettlement of man (the worker) at the center of industrial processes. The concept of Industry 5.0 requires the use of advanced technologies to support man in his actions, to help him to progress and to offer him the solutions for his needs and interests.

Industry 5.0, in order not to repeat the mistakes/omissions of Industry 4.0, it wants to be an open and continuously evolving concept (European Commission, 2021).

In order to better understand the context in which the two concepts—Industry 4.0 and Industry 5.0—appeared and work, we resorted to a SWOT analysis (Fig. 4).

Strengths (advantages)	Opportunities
✓ Capitalizing on the results of technological progress ✓ Internet of Things ✓ Digital applications for manufacturing ✓ Increasing the productivity, the volume and the range of production ✓ Advanced globalization (economic and social)	✓ Increasing the consumption of products and services ✓ Demand for new products/services ✓ Consumer interest in products with a high degree of technology ✓ Promotion and development of technologies using clean energy
INDUSTRY 4.0	
Weaknesses (disadvantages)	**Risks**
✓ Accentuation of social inequality ✓ The deepening of regional discrepancies regarding theeconomic development ✓ Discrepancies between the producers involved in the same manufacturing cycle ✓ Fragmentation of industrial domains ✓ Fragility of critical infrastructures ✓ Excessive use of natural resources ✓ Environmental pollution	✓ Limited digital skills ✓ Reduced investment capacity ✓ Exhaustion of resources ✓ Social dissatisfaction ✓ Disruptive factors, such as: pandemics, wars

Fig. 4 Comparative SWOT analysis for Industry 4.0, respectively Industry 5.0 (authors' proposal)

Strengths (advantages)	Opportunities
✓ Technological strategies (co-operation between the scientific environment, the business environment and decision-makers)	✓ Rethinking theEuropean and world policies regarding the original principles of Industry 4.0
✓ High degree of robustness for industry and production flexibility	✓ Development of new strategies regarding the implementation of digital technologies and economy
✓ Safety in the operation of critical infrastructures during periods of crisis (see the period of the Covid-19 pandemic and the crisis generated by the war in Ukraine)	✓ Reduction of energy and raw material consumption
✓ Man-machine cooperation. The development of collaborative artificial intelligence	✓ Large-scale application of reuse and recycling
✓ The stability of the employed staff (as a labor resource and as a level of well-being)	
✓ Environmental protection (carbon-neutral industry - "green production")	
✓ Energy efficiency (both in the field of production and also in the field of consumption)	
✓ Promoting the circular economy	

Fig. 4 (continued)

INDUSTRY5.0	
Weaknesses (disadvantages)	**Risks**
✓ Constraints on scientific research, manufacturing and consumption ✓ The deepening of regional discrepancies regarding the economic development, against the background of unequal access to sustainable resources ✓ Fragmentation of industrial domains ✓ Relatively high costs for implementing ecological solutions for manufacturing and consumption ✓ Inducing a sense of crisis in society	✓ Threats to fundamental human rights ✓ Diminishing the role of democracy and minimizing the social values ✓ Imbalances in the human-machine relationship, in favor of the machine (artificial intelligence) ✓ Exaggerated energy consumption required by the new applications

Fig. 4 (continued)

According to (European Commission, 2021), Industry 5.0 recognizes the power of industry to achieve societal goals beyond jobs and growth, to become a resilient provider of prosperity, making production respect the limits of our planet and placing the well-being of the industrial worker at the center of the production process.

The question from which we started in carrying out this study—INDUSTRY 5.0 represents a completely innovative concept or is it, in fact, a complement to the INDUSTRY 4.0 concept?—it can be maintained as a subject of analysis because, depending on the context, we can consider both answers to be correct, thus:

- **Industry 5.0 is an innovative concept** because, until this moment, the industrial revolutions had their objectives oriented, in general, to the issue of technologies, technical progress and the integration of scientific research with manufacturing. It is for the first time that an industrial revolution refers to the social factor as an element of interest in the formulation of economic and environmental policies;

- **Industry 5.0 is a complement to the concept of Industry 4.0** because, upon closer analysis, we find that, trying to answer the specific questions of Industry 4.0 and benefiting from similar advantages, it aims to avoid the limits of the previous revolution. In this context, Industry 5.0 has identified this "historical" omission of the industrial revolutions, namely the social factor, an aspect which it extends beyond the organization of production or consumption, to the point of proposing a collaborative relationship-based on definitive social laws—between man and machine, to man's advantage.

3 Sustainability—Basic Principle of Industry 5.0

When discussing the sustainability of an activity or a process, reference is made to its ability to be carried out for as long as possible, under the conditions of the conservation of integral natural resources. The realities of the last decades have forced us to develop the classic concept of sustainability, assigning to industrial systems restrictions regarding the use and development of natural resources under the conditions of maintaining the balance of the environment, so without compromising the possibilities of meeting the needs of future generations.

In this context we cannot—and must not—avoid discussing the sustainability of the industrial sector and its conditioning on the sustainability of the energy resources of the planet. Industry 5.0, in a first analysis, seems to be similar to the concept of Industry 4.0 that has received a facelift. But things are deeper, why (see Fig. 2):

- because, beyond the economic development based on the results of scientific research, beyond the response to society's consumption requirements and beyond the intentions—most of the time remaining at the declarative level—of the factors responsible for protecting the natural environment, Industry 5.0 gives to the society a great responsibility. This responsibility refers to being balanced—we, the people—in formulating the requirements and in establishing the assortment range and the volume of consumption;
- because Industry 5.0 brings social issues to the fore, trying to impose a balance within human communities and, why not, at the European and global level. Under this aspect, the problems to be solved are numerous, serious and involve consensus at the micro- and macro-economic level. We must understand that man is part of the natural environment and, consequently, at every social level, we must apply measures to protect man—workers and consumers—so as to result in a sustainable humanity;
- because, beyond the measures to protect the environment—also identifiable in the concept of Industry 4.0—Industry 5.0 identifies the need to ensure technological security, the safety of critical infrastructures and the development of alternative energy solutions with the assumption—at a macro-economic and macro-social level—of the "rationalization" (that is, the manifestation of a reasonable demand in accordance with decent needs) of the demand.

Probably, in the development of this new vision on the meaning of industrial development, an important role was played by the two important disruptive factors that affected our trajectory—as humanity—in the last 3–4 years: the Covid 19 pandemic crisis and the escalation of the war in Ukraine. These two events require balanced solutions targeting—exactly as the Industry 5.0 concept wants to declare—the human society (solving territorial, social and public interaction problems) and the natural environment (solving problems related to natural resources, their access, use and conservation).

Since the current era is, we can say, the result of a major industrial revolution, it follows that the sustainability of humanity involves, among other things, ensuring the sustainability of industry. Continuing the reasoning, we must bear in mind that the industry is a large consumer of energy resources and, at the same time, a large producer of harmful products for the environment, if we refer to the use of fossil fuels, especially. This shows us that energy sustainability offers an additional guarantee to industrial sustainability, just as the Industry 5.0 concept implies.

Even if, at a summary analysis, the investment in renewable energy seems unprofitable, however, efforts must be increased in the direction of the development and use of non-polluting energy resources, as they offer a series of advantages in supporting the approach of sustainable industry. Among these advantages we mention (Chhabra, 2021):

- the use of renewable/sustainable energy in the industrial sector is that it significantly reduces the total carbon footprint of industries;
- in a consumer society, where the buyer's requirements are the main priority of the industry, the transition to sustainable energy will help the environment and, at the same time, highlight the futuristic industrial results, increasing the degree of appreciation of the products;
- the industry using green energies will reduce mechanization from polluting energy sources and will require the development of new occupations/specializations creating more job opportunities and stimulating the entire economic and social system.

The conclusion that emerges is that Industry 5.0 embraces the idea of a sustainable industry and opens the way to a healthier future, the reference point being considered—and here comes the novelty in relation to Industry 4.0—"a totally sustainable society".

4 The Energy Role in the Current Economic Context

If Industry 4.0 created new innovative perspectives for the industrial branches and thereby contributed to the increase in production, the next step is Industry 5.0, that of taking automation to a higher level both by increasing the efficiency of technological processes and operations, as well as by reducing the size workforce and energy consumption (Akkaya & Tabak, 2022; Traşcă et al., 2019).

From a conceptual point of view, Industry 5.0, the future stage of the industrial economy refers to the integration into a unitary whole of products, processes, machines, software and industrial robots for the realization of Industry 5.0. This multi-criteria integration involves the dual integration of human intelligence with machine intelligence and monitors and analyzes the results using the Industrial Internet of Things (IIoT) and artificial intelligence (AI) (Akkaya, 2021; Akkaya & Tabak, 2022). In order to speed up the manufacturing process and economic efficiency, the creation of a new generation of robots called Collaborative Robots (Cobots) is being explored. Also in this framework, it is necessary to reduce the consumption of material resources, energy and labor through the optimal design of the products and the correct choice of the manufacturing process, offering customers more customized and personalized products.

Taking into account the current political social economic framework and the fact that Industry 5.0 is based, among other things, on the reduction of energy consumption, it can be stated that energy is a very important factor that influences the achievement of this objective in good conditions.

Considering the multitude of events in 2021, it can be said with certainty that they have negatively influenced the industry and the energy market. The efforts made by specialists in the energy fields globally, to reduce greenhouse gas emissions in accordance with the desired Net Zero, have registered a considerable increase (Hoinaru et al., 2019). Although it is possible that for everything that was discussed in Glasgow at the 26th UN Climate Change Conference of the Parties (COP26), no consensus was reached, nevertheless important commitments and concessions regarding climate change resulted. Also, independent of government policies, many industrial companies have taken important steps to reduce polluting emissions to achieve the desired Net Zero, thus paving the way for other entities in common industrial fields. This is due to the fact that the transition to a clean industrial production is desired, but in many cases there have been pressure actions from both public and private investors, as well as from shareholders and financiers.

The economic perspectives of the European Union, affected by the increase in inflation and the energy crisis, have become more and more precarious, with very high risks that can lead to recession.

Even though at the global level, the significant reduction of greenhouse gas emissions has been proposed as an objective to mitigate climate change, and both governments and companies in the energy fields want to switch from the use of fossil fuels and invest in renewable energy sources, the current context of the energy crisis inhibits the rapid transition to energy production technologies from sustainable sources (Adebayo et al., 2022; Fernández-González et al., 2020, 2021).

In the field of energy technologies, some emerging trends have been outlined for the year 2022 and for the following periods.

1. The field of the energy industry is based on the desired Net Zero.
2. Special attention given to technological systems for storing various forms of energy.
3. Hydrogen, the energy source of the future (Neacşa et al., 2022a).

Specific technologies for obtaining energy have a very important role in the industrial sustainability (including industry 5.0 sustainability) and, in general, in the sustainability of our planet by reducing greenhouse gas emissions, especially carbon dioxide.

Technologies that have an upward trend, Carbon Capture and Storage, Direct Air Capture, systems for energy storage and technologies that use hydrogen, are the innovative energy solutions both for the year 2022 and for future periods (Neacsa et al., 2020; Neacşa et al., 2022a).

Considering the trends of continuous increase in energy prices and disruptions in energy networks, at the level of the European Union there is an increased interest and continuous debates on how to mitigate their negative effects and reduce dependence on fossil fuels and natural gases from Russia.

The decrease or blocking of the supply of energy and natural gas from Russia to Europe has led to both rising costs and energy insecurity for the economies of many European countries.

The price of energy has direct effects on exports, imports and implicitly on inflation, situations that in turn produce major implications in the current account, the trade balance and, therefore, in the economy. In the same context, prices for different forms of energy affect spending, taxes and the budget deficit, as well as public debt, affecting the economy in turn. Also, production costs affect energy prices, which in turn directly influence exports, commercial activities and implicitly the economy (Neacsa et al., 2020). All these situations can act at the same time producing a more accentuated and combined effect on the economy.

Consumers in the European Union are now facing the challenge of rising cost of living and falling real purchasing power, which has affected their consumer confidence.

At the same time, the increase in the cost of energy had a negative impact on industrial production, which decreased dramatically, ultimately leading to the closure of some capacities of large energy-consuming companies, producing more negative effects on the European economy.

Mankind has experienced several energy transitions over time as new resources have been discovered and technological inventions have been made to facilitate the production of various forms of energy.

The new trend towards a just transition from conventional (fossil) to alternative renewable energy sources is identified as a complex, multi-sided process, which is driven both by the limits of oil and gas reserves and the need to use resources that have little or no environmental impact. Population growth, urbanization, increased economic activity and globalization are some of the factors that have led to increased energy consumption with devastating effects on the environment (Neacsa et al., 2022b, 2022c; Embassy of Algeria, 2021; Noja et al., 2022). The current energy transition is therefore a politically driven process, as countries around the world have realized the need to take concrete steps to protect the environment (Neacsa et al., 2022b, 2022c; Embassy of Algeria 2021; Erokhin & Tianming, 2022). However, the energy transition comes with a number of economic, technical, social and energy security challenges. In addition, the main opportunities and challenges generated by the energy transition for different stakeholders in the process are presented (Neacsa et al., 2022b, 2022c; Embassy of Algeria, 2021). The analyses undertaken in this sensitive field demonstrate the complexity of the phenomenon, its multidimensional character and the importance of the involvement of public authorities and international institutions in the energy transition process (Neacsa et al., 2022b, 2022c; Embassy of Algeria, 2021).

In the current socio-political-economic context, in Europe, the population (domestic and industrial consumers) is increasingly affected by the very high values for the payment of utility bills, in this case those for electricity and natural gas. In the last period (first half of 2022) consumers have received notices about consistent price changes (increases).

There are two ways of trying to explain this upward slope of prices for accessing different energies:

- The first variant has as its starting point the fact that the political class bears a large part of the blame because it is incompetent by allowing suppliers through legislation, with the rest of the blame, an unjustified increase in prices for accessing different forms of energy; there are also consumers who, in addition to the political class, find suppliers, distributors or transporters guilty;
- The second variant claims external causes as factors of influence: the post-epidemic situation caused by COVID 19 SARS 2 and that caused by the special operation carried out by Russia in Ukraine (everywhere in the world the same happens, implicitly also in Europe).

The two approaches are partially true, but the current situation is much more complicated, as many more causes have been identified that generated this critical situation from a socio-economic point of view. As a result of an analysis of the socio-economic framework, of the energy situation, it can be highlighted that many of the causes, objective and/or subjective, remain unknown to the population. The implications of these causes also remain hidden.

There are a lot of misinformation and fact manipulation actions for which it can be said that public opinion is marked by feelings of confusion (natural up to a point).

During the most recent meeting, which had as its subject the Composite Main Indicators (CMI), of the Organization for Economic Cooperation and Development (OECD), a statement was issued stating that being negatively influenced by the historically high inflation values, by the low confidence of consumers in the economy and the decrease in stock market values, the CMI indices (Data extracted from https://data. oecd.org/leadind/composite-leading-indicator-cli.htm on 09/09/2022) from remain below the normal level and continue to anticipate a downward trend at the level of the large OECD economies (see Fig. 5—Amplitude adjusted, Long-term average = 100, Aug 2010–Aug 2022) (National Action Plans 2022).

According to the report issued by the OECD, these indices, which are specifically designed to anticipate specific turning points in economic activity for a future time horizon of six to nine months, continue to indicate a pessimistic trend for the outlook in most major economies.

Fig. 5 Evolution of CMI

A vital response to the Sustainable Development Goals (SDGs) is the transition to renewable energy, increased efficiency and energy conservation. In terms of energy sources, this transition will also have a major impact on the economies of the European Union by creating independence from fossil fuel sources.

As a general conclusion regarding the current energy context, it can be stated that the current socio-political framework is clearly a new energy transition, which is clearly different from the others in terms of its pace but also in terms of the main driving force behind it, namely the international and EU institutions. This energy transition is the first process of its kind that is politically driven and not generated by a natural evolution of the world economy and humanity. The reason for the need to adopt concerted solutions at EU level is to take into account the major risks that global warming may generate for humanity.

The European Union has both considerable renewable potential, given the sunshine duration and wind speed, and a large scope for action to increase energy efficiency in the residential building sector. However, existing public policies continue to ignore the benefits that these actions can bring in reducing the energy vulnerability of the EU economies (Neacsa et al., 2022b, 2022c).

The energy transition offers solutions to protect the environment, but it raises economic, social and technical issues that companies, public authorities, financial institutions and researchers need to address. The need to protect the environment and the decline in fossil fuel resources have triggered a new energy transition, with renewable energy being the new type of energy that will ensure the transition to a low-carbon economy. The energy transition is accompanied by a complex process of changing attitudes and behaviours of energy consumers, producers and investors. The consequences are far-reaching not only in economic and environmental terms, but also in social terms, with renewable energy as a solution for reducing energy poverty and developing rural communities. Changes in consumer attitudes and the metamorphosis of business strategies are observable in all countries, with the energy transition being a reality even in the financial sector. Adaptation but also innovation are the watchwords for all categories of stakeholders. The complexity of the energy transition phenomenon requires the involvement of companies, consumers, portfolio investors and the education system, which must encourage a change of mentality and improved behavior.

Decarbonisation of the European economy is a bold objective set by the European Green Pact, with the transition to clean energy being one of the most important directions for action in EU countries. Through concrete measures, EU countries aim to interconnect energy systems and integrate renewable energy resources, increase energy efficiency, design green products, fight economy vulnerability. Access to affordable energy in a secure, sustainable and modern way (Sustainable Development Goal, SDG7) is the solution to combating economy vulnerability, a phenomenon that is manifest in both developed and emerging economies.

Consumers in the European Union are now facing the challenge of rising cost of living and falling real purchasing power, which has affected their consumer confidence.

At the same time, the increase in the cost of energy had a negative impact on industrial production, which decreased dramatically, ultimately leading to the closure of some capacities of large energy-consuming companies, producing more negative effects on the European economy.

5 Energy in the Context of Industry 5.0

Industry 5.0 aims to simulate the attractiveness and competitiveness of industries based on the improvement of working environments, large-scale technology and the design of production systems adaptable to crisis situations in particular (Keidanren, 2018).

Industry 5.0 appeals—we can say excessively—to emerging technologies such as: blockchain, digital twins, artificial intelligence and machine learning (AI/ML), edge and IoT, augmented reality/virtual reality (AR/VR) (GE Digital, 2022).

The emerging technologies will help to (GE Digital, 2022):

- monitoring, processing and interpretation of industry data;
- predictive modeling;
- increasing the efficiency in the management of manufacturing processes and the operation of facilities;
- environmental monitoring;
- resource management and supply efficiency.

Industry 5.0, introducing the principle of human-centeredness, substantially changes the coordinates of scientific research activity in the laboratory. Changes in research activity were also noted within Industry 4.0, which required the adaptation of the research project to real production conditions and manufacturing capabilities.

So, Industry 4.0 is practically not related to the humans. It can be considered that Industry 5.0 somehow complicates the laboratory activity by forcing research—but also manufacturing—to support man and his interests.

When discussing about the human—in the opinion of the authors, by referring to the principles of Industry 5.0—not only the working personnel should be taken into account, but also the beneficiaries of the products and services. Perhaps, in most specialist analyzes on this topic, beneficiaries are not mentioned, this is probably due to a prior customer-centricity imposed by management systems—especially quality management systems—that already enjoy experience in applications.

The sustainability and the resilience are the two principles of Industry 5.0 that force the provision of industrial solutions that are environmentally friendly and, at the same time, adaptable to current conditions. This is not easy to ensure. We need to think of policies and strategies that go beyond even the boundary of recovery and use of waste or the processes of acquisition and use of green energy.

Emerging technologies even if we intend 'to use them as a priority for the development of environmentally friendly industrial solutions, they are not always sustainable in all the applications. AI/ML and blockchain, among others, are computationally intensive. The International Energy Agency (IEA) reported that bitcoin alone consumes more than 100 TWh (terawatt-hours) per year, which is equivalent to the annual electricity consumption of the Netherlands. As a specific example, edge and IoT devices distribute the computer's carbon footprint to the edges of the network. They also contribute to the generation of electronic waste (GE Digital, 2022; Forrester Research, 2022).

Economic, industrial and environmental policies inevitably collide with the problem of energy production and consumption (Fig. 6). The 5th industrial revolution requires, more than ever, the identification of new, non-polluting, renewable, inexhaustible energy solutions. Finding the balance between material well-being and sustainable society—including here also the environment in which the society manifests and develops—is a challenge whose solution may require major sacrifices or perhaps even a return in time from a behavioral point of view.

The risks accompanying the emerging digital technologies include (GE Digital, 2022; Forrester Research, 2022):

- increasing the power, water and cooling requirements that enhance the greenhouse gas emissions;
- expanded manufacturing risks and resource requirements for new chips, devices and robots;
- increasing the volume of electronic waste and toxic chemicals.

It follows that the solutions offered by the emerging technologies as well as the use of renewable energy sources (wind and solar) are not sufficient in the context of modern society. Cutting-edge technologies and applications are not as harmless from an energy point of view as one might think. IT equipment requires considerable energy resources (for manufacturing, operation and neutralization after use). The increase in operating capacity implies a directly proportional increase in energy consumption, if we refer only to operation and cooling.

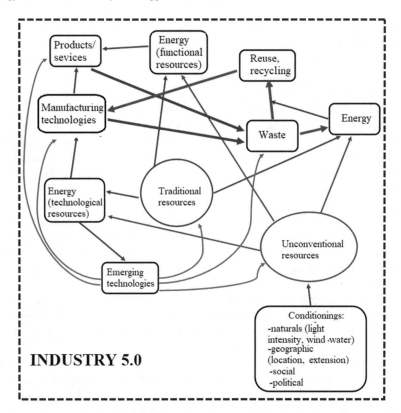

Fig. 6 Energy challenges of Industry 5.0

Also, the expansion of applications in terms of electricity-based transport involves increasing the production and distribution capacities of electrical energy far beyond the current, real possibilities of electrical systems.

The overloading of the energy systems risks interruption in the supply of some of the economic and/or social structures (supplying household customers, the banking system, critical infrastructures).

Green energy sources, unfortunately, depend on natural factors and do not offer sufficient stability in operation: the wind power plants are dependent on the wind speed values; the photovoltaic cells have maximum efficiency in a certain range of light intensity; the hydroelectric plants—which can be considered as the most reliable from a technical point of view—are put in difficulty due to the radical climate changes.

The circular economy, even in the context of Industry 4.0, has only partially offered solutions regarding the resource economy. Waste recycling responds, in particular, to the request to save material resources, but recycling technologies are, in turn, energy-consuming and, in certain situations, even polluting.

If we refer to the critical infrastructures, even if they enjoy priority in ensuring energy resources—which, as seen, are limited—disruptive factors can intervene—such as the example of the war in Ukraine—which may disrupt the continuity of operation or respecting their optimal parameters.

We have only referred to energy sources and their important applications. However, it is worth mentioning the food resources of mankind which constitute the engine of the functioning of the human body and which, in turn, are dependent, from an industrial point of view, on energy resources. And in this case, beyond the uneven geographical distribution, the dependence on environmental conditions in continuous degradation, disturbing factors such as war situations in large food producing areas also intervene.

It is the task of the 5th industrial revolution to identify solutions to all these problems. From an energy point of view, the future will probably be the hydrogen. Efforts must be intensified for the development of efficient technologies for the production, transport and storage of hydrogen, as well as the adaptation of current consumers to use it, at least partially, as a fuel, as well as the development of consumers fully adapted to the new fuel.

The traditional techniques regarding the protection and preservation of the environment should not be neglected either. Here we mention only the afforestation activities that should be a global priority, with beneficial effects on agriculture, air quality, returning to a normal climate.

The issue of sustainability as a feature of Industry 5.0 is, as seen, inextricably linked to obtaining, using and conserving energy. We try, in this sense, a modeling of the industrial problem from the point of view of sustainability, representing, definitively, the measure of the reliability of the industrial system.

6   Conclusions

The evolution of technologies, especially in the modern era, requires important resources, including energy.

The nature has limited resources, which are unevenly distributed on the Earth's surface. Throughout history, the rush for resources has led to serious conflicts that have often substantially marked the human society.

The modernization of the human habitat has led to the development of some of the most sophisticated technological solutions that meet consumer requirements, reaching the stage of product customization (see also Industry 4.0).

The time has come when humanity must restore the balance between needs and resources. Beyond restoring the fragile natural balance, Industry 5.0 tries to offer solutions for restoring, including, the social balance. Future concerns will be directed towards facilitating reasonable access to resources, subject to their efficient use.

Industry 5.0 draws attention to the need to stop excesses both in consumption and in social relations, within productive systems and within society as a whole.

Industry 5.0 puts the sustainability of humanity at the expense of solving social problems and respecting the human rights.

References

Adebayo, T. S., Oladipupo, S. D., Adeshola, I., & Rjoub, H. (2022). Wavelet analysis of impact of renewable energy consumption and technological innovation on CO_2 emissions: Evidence from Portugal. *Environmental Science and Pollution Research, 29*(16), 23887–23904.

Akkaya, B. (2021). Leadership 5.0 in Industry 4.0: Leadership in perspective of organizational agility. In *Research Anthology on Cross-Industry Challenges of Industry 4.0* (pp. 1489–1507). IGI Global.

Akkaya, B., & Tabak, A. (2022). Leader robots (LRs): The future managers of digital organizations. In *Agile Management and VUCA-RR: Opportunities and Threats in Industry 4.0 Towards Society 5.0* (pp. 215–222). Emerald Publishing Limited.

Chhabra, A. (2021). Sustainable energy for industry: Navigating the world of sustainability for a better future. https://blog.se.com/mining-metalsminerals/2021/11/03/sustainable-energy-for-industry-navigating-the-world-of-sustainability-for-a-better-future/

Erokhin, V., & Tianming, G. (2022). Renewable energy as a promising venue for China-Russia collaboration. In *Energy Transition* (pp. 73–101). Springer.

European Commission. (2021). Directorate-general for research and innovation. In M. Breque, L. De Nul, & A. Petridis, *Industry 5.0: Towards a sustainable, human-centric and resilient European industry*. Publications.

Felsberger, A., & Reiner, G. (2020). Sustainable Industry 4.0 in production and operations management: A systematic literature review. *Sustainability, 12*, 7982.

Fernández-González, R., Arce, E., & Garza-Gil, D. (2021). How political decisions affect the economy of a sector: The example of photovoltaic energy in Spain. *Energy Reports, 7*, 2940–2949.

Fernández-González, R., Suárez-García, A., Alvarez Feijoo, M. A., Arce, E., & Díez-Mediavilla, M. (2020). Spanish photovoltaic solar energy: Institutional change, financial effects, and the business sector. *Sustainability, 12*(5), 1892.

Grabowska, S., Saniuk, S., & Gajdzik, B. (2022). Industry 5.0: Improving humanization and sustainability of Industry 4.0 (springer.com).

Hoinaru, R., Negreanu, A., & De Luca, A. (2019). *Driving down greenhouse gases: A roadmap for the Paris Agreement*. European Parliament House Publishing, (forthcoming) Search in.

https://algerianembassy-japan.jp/en/2021/06/18/renewable-energy-algeria-has-africas-third-largest-installed-capacity-in-2020/. 09.09.2022

https://www.imt.ro/TGEPLAT/eveniment_9.10.2017/Prezentari/Prezentare%20Generala%20TGE-PLAT-9%20Oct%202017%20RM.pdf/. Accessed Sept 20, 2022

https://globalnaps.org/issue/sustainable-development/. 09.09.2022

https://www.keidanren.or.jp/en/policy/2018/085_overview.pdf. Accessed Sept 15, 2022

https://www.ge.com/digital/blog/digital-oxygen-energy-transition. Accessed Sept 15, 2022

https://www.ge.com/digital/sites/default/files/download_assets/forrester-sustainability-profitability-q-and-a.pdf. Accessed Sept 30, 2022

Neacsa, A., Panait, M., Muresan, J. D., & Voica, M. C. (2020). Energy poverty in European Union: Assessment difficulties, effects on the quality of life, mitigation measures. Some evidences from Romania. *Sustainability, 12*, 4036. https://doi.org/10.3390/su12104036

Neacșa, A., Panait, M., Mureșan, J. D., Voica, M. C., & Manta, O. (2022a, March 4). The energy transition between desideratum and challenge: Are cogeneration and trigeneration the best solution? *International Journal of Environmental Research and Public Health, 19*(5), 3039. PMID: 35270731; PMCID: PMC8910140. https://doi.org/10.3390/ijerph19053039

Neacsa, A., Rehman Khan, S. A., Panait, M., & Apostu, S. A. (2022b). *The transition to renewable energy—A sustainability issue? Energy transition—Economic, social and environmental dimensions* Chapter 2, August 30, 2022b, Industrial Ecology Book Series (IE), Springer Book.

Neacsa, A., Eparu, C. N., & Stoica, D. B. (2022c). Hydrogen–natural gas blending in distribution systems—An energy, economic, and environmental assessment. *Energies, 15*, 6143. https://doi.org/10.3390/en15176143

Noja, G. G., Cristea, M., Panait, M., Trif, S. M., & Ponea, C. Ș. (2022). The impact of energy innovations and environmental performance on the sustainable development of the EU countries in a globalized digital economy. *Frontiers in Environmental Science, 777*.

Romero, D., Bernus, P., Noran, O., Stahre, J., & Fast-Berglund, Å. (2016). The operator 4.0: Human cyberphysical systems & adaptive automation towards human-automation symbiosis work systems. In *IFIP International Conference on Advances in Production Management Systems* (pp. 677–686). Springer.

Trașcă, D. L., Ștefan, G. M., Sahlian, D. N., Hoinaru, R., & Șerban-Oprescu, G. L. (2019). Digitalization and business activity. The struggle to catch up in CEE countries. *Sustainability, 11*(8), 2204.

Assessing the Drivers Behind Innovative and Creative Companies. The Importance of Knowledge Transfer in the Field of Industry 5.0

Carlos Rodríguez-Garcia, Fernando León-Mateos, Lucas López-Manuel, and Antonio Sartal

Abstract Knowledge-intensive entrepreneurship (KIE) firms are widely recognized in the literature as the type of innovative companies that can exert the greatest influence on a region's economic and social development. This has sparked a growing interest among governments and policymakers in developing initiatives to enhance the various drivers behind this type of entrepreneurship, although the results are not always desirable. Accordingly, we proposed a case study comparing the two regions comprising the Galicia-North Portugal Euroregion to assess the key driver(s) behind successful KIE firms development during the period of 2015–2019. The data comparison of both regions points to a greater relevance of regional knowledge transfer to develop this type of organizations. Thus, the exploratory results of this analysis should encourage public administrations to place special emphasis on initiatives aimed at transferring knowledge among the various agents involved in the innovation system. This would not only help them to design more efficient public policies for promoting Industry 5.0 but also it would help to enhance regions' development in the long term.

Keywords Knowledge-intensive entrepreneurship firms · Innovative and creative companies · Regional knowledge transfer · Entrepreneurship policies · Galicia-North Portugal Euroregion · Case study

1 Introduction

Ever since Schumpeter (1934) asseÜrted that entrepreneurship drives economic development, academics and policymakers have been racing to unravel how and when entrepreneurs' activities can create a disruptive force in the economy that can, in turn, lead to economic growth (Kyllingstad, 2021; Malerba & McKelvey, 2020). After all, economic growth is one of the main objectives of any government, as it is directly related to job creation, higher tax revenue and ultimately higher standards of living for citizens (Aparicio et al., 2021; Veenhoven & Vergunst, 2014).

C. Rodríguez-Garcia · F. León-Mateos · L. López-Manuel · A. Sartal (✉)
School of Economics and Business, University of Vigo, Vigo, Spain
e-mail: antoniosartal@uvigo.es

© The Author(s), under exclusive license to Springer Nature Switzerland AG 2023
C. F. Machado and J. P. Davim (eds.), *Industry 5.0*,
https://doi.org/10.1007/978-3-031-26232-6_5

91

Many authors have established that entrepreneurship and innovative firms have a positive effect on a region's or country's industrial development and economic growth (e.g., Urbano & Aparicio, 2016; Achim et al., 2021). However, ample evidence in the literature also points to the fact that not all types of new firms are equally relevant for achieving economic growth (Nightingale & Coad, 2014). A growing number of authors advocate the need for the knowledge-intensive entrepreneurship (KIE) firms, which link the production of new technological knowledge to its eventual commercialization. These are the type of new firms, innovative and creative companies, with real potential for promoting regions' industrial and social development (Malerba & McKelvey, 2020).

Authors such as Szerb et al. (2019) and Nicotra et al. (2018) go further and highlight not only the positive impact of KIE firms on employment but also their link to sustainable and inclusive development. In fact, the emphasis placed on this type of knowledge is increasingly palpable in the development of public policies in the field of Industry 5.0. Government initiatives designed to foster entrepreneurship in general, and KIE in particular, have become increasingly common in recent years. The Build to Scale Program that the U.S. Economic Development Administration (2021) developed, Innovative Solutions Canada (Government of Canada, 2021) and the Entrepreneurs' Programme that the Australian Government (2021) promotes are just some of the best-known initiatives in this regard. However, it is precisely at this point that our research issue arises.

Although many authors describe the need for and the relevance of KIE on a theoretical level (e.g., Malerba & McKelvey, 2020; Stam & van de Ven, 2019), the public initiatives in this field does not always seem to achieve the levels of success expected (Caloghirou & Llerena, 2015). Public initiatives, such as Technium and High Performance Computing (Wales), the Intermediate Technology Initiative (Scotland) or the Research and Development (R&D) Promotion Programs of the Economic Development Administration (USA), are clear examples of failed actions (Pugh et al., 2018). In addition, the European Commission has directly discouraged certain programs, such as BRUSTART (Belgium), TechInvest (UK) and the Connect startups platform (Poland), due to their low levels of effectiveness (European Commission, 2021). A common problem with these programs is that they have led to limited growth in the number of new firms 5.0 created. In several of the programs, such growth amounted to just one-third of the objectives set (Pugh et al., 2018). However, the main problem lies in the low return obtained from public resources, which are already scarce and very much needed in many other areas (Kasabov, 2016). In view of this situation, several authors have tried to identify the key factors favoring both the emergence of KIE firms, i.e., innovative, and creative firms, and its subsequent development (e.g., Stam & van de Ven, 2019; Malerba & McKelvey, 2020).

The abovementioned studies have helped us to pinpoint the main drivers behind KIE firms. Most of them identify, in an aggregated manner, the various drivers involved in the development of this type of entrepreneurship, providing general recommendations for KIE's development and implementation. However, these "generalist prescriptions" do not delve into each driver's specific influence on KIE's emergence and development. In fact, the reason why certain KIE promotion initiatives are so successful (e.g., the Build to Scale Program or Innovative Solutions Canada) whereas others fail (e.g., Technium or High Performance Computing) is because these global analyses, which consider all drivers equally, do not identify the drivers that could be truly key for this type of entrepreneurship. In fact, public incentives, if they are to be more efficient, should focus on incentivizing only those drivers that truly lead to the development of KIE firms in the field of Industry 5.0.

With this idea in mind, our work is aimed at separately assessing the importance of the key drivers behind KIE development by means of a case study in the Galicia-North Portugal Euroregion. First, we conducted a systematic literature review (SLR) with the aim of identifying the main drivers behind KIE. We then proposed an exploratory analysis through a case study comparing the two regions comprising the Galicia-North Portugal (GNP) Euroregion.

This analysis was possible because the socioeconomic and industrial contexts of these two sub-regions are practically identical, and nevertheless, Northern Portugal has a significantly higher level of innovative companies. This situation encouraged us to delve deeper into the possible causes of this higher KIE performance with the intention of generalizing the knowledge obtained. This exploratory analysis, based on public and robust data from various official databases, provides an interesting starting point that can guide regional policymakers in the efficient allocation of public funds to promote successful innovative initiatives 5.0 in the industrial field.

2   Drivers Behind KIE Firms: The Key Role of Knowledge Transfer

2.1   Systematic Review of the Literature: The Main Drivers Behind KIE

The first step of our research was an SLR aimed at identifying the main drivers behind KIE. This method not only provides the opportunity to map all of the knowledge gathered in this field of study but also is key for trying to understand and conceptualize it (e.g., Gao et al., 2019). Following Tranfield et al. (2003), we conducted this SLR in three stages: (i) planning the review, (ii) conducting the review and (iii) reviewing the findings.

(i) **Planning the review.** Having ascertained that no pre-existing SLR existed on the drivers behind KIE firms, we defined the SLR's structure. In line with previous SLRs (Foss & Saebi, 2017; Sivarajah et al., 2017), we decided to use the two main scientific search engines, Web of Science (WoS) and Scopus (SJR). To specify the conceptual boundaries, we selected search equations that included the keywords (e.g., KIE, drivers, enablers, etc.) as set out in Fig. 1. In addition, we refined and reviewed all articles and matching technical articles to ensure that they aligned with our objectives. Specifically, we checked to make sure that the term "KIE" was included in the articles' titles, abstracts or keywords. We selected articles published in English to ensure readability. An additional quality criterion was adopted: including only articles published in journals indexed in quartiles 1 and 2 of the WoS and SJR 2020.

(ii) **Conducting the review**. The search stage took place between July and August 2021. The search and selection processes were similar in both cases (WoS and SJR). We located the main empirical works of the past decade (2011–2021) using search equations that related the terms "knowledge intensive entrepreneurship," "drivers," "enablers," "factors," "dimensions," "background," "entrepreneurial ecosystem" and "measurement" using Boolean operators. As the study of KIE is a topical issue in the literature, the review focused on the past decade.

The searches focused on titles, abstracts and keywords, and the categories selected included management, business, economic and the social sciences (in the case of Scopus, accounting, econometrics and finance were also included). The first selection was refined by eliminating duplicate articles and eliminating those not written in English. For the purpose of further ensuring objectivity, only papers published in indexed WoS and SJR journals were included. Thus, conference reviews, book reviews, book chapters and undefined research papers were excluded to maintain the quality of the study.

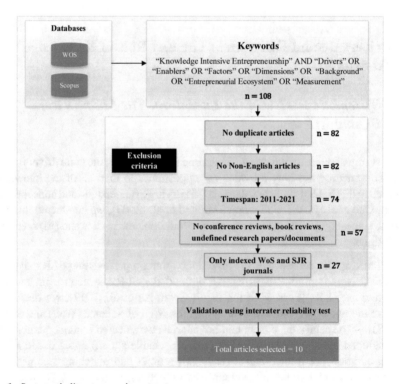

Fig. 1 Systematic literature review structure

(iii) **Reviewing the findings.** Finally, we used an inter-rate reliability test to eliminate any articles that did not fit our search objectives. The key criteria for inclusion were that KIE had to appear as the centrepiece of the article and that it had to be related to drivers that could help it to develop. To maximize the robustness of the study, the authors independently decoded the selected articles, and the findings were subsequently compared to assess possible differences. All articles were found to have minimal or no differences in scores, ensuring the quality of the review. Our systematic literature review provides an analysis of 10 key papers (Fig. 2). Additionally, we include the entrepreneurship drivers proposed by the Global Entrepreneurship Monitor (GEM, 2017a, 2017b). The main dimensions selected in our literature review do not differ from those that this body has highlighted.

2.2   Identification of KIE Drivers in the Field of Industry 5.0

The SLR allowed us to conclude, as expected, that a broad consensus exists on the parameters of the socioeconomic environment that can influence a region's or country's KIE (e.g., Stam & van de Ven, 2019). Without forgetting the importance of each individual entrepreneur's attributes, the papers analysed highlight five drivers in the environment as key factors behind KIE's development in a region: i) institutional quality, ii) industrial specialisation, iii) social capital, iv) human capital and v) knowledge transfer (see Fig. 2).

In line with the concept map we have proposed, we describe below the various drivers identified in the review. However, we leave the "knowledge transfer" driver, which appears in all of the studies and gives rise to our proposition, until the end. We also stress the fact that the other drivers explicitly or implicitly reflect the need for this facilitator to accompany them for them to have a significant impact on the development of KIE firms.

(i) **Institutional quality.** Institutions are among the main sources of incentives (and disincentives) for entrepreneurial activity, as they facilitate reliable interactions among the various agents involved in the entrepreneurial process (Malerba & McKelvey, 2020; Szerb et al., 2019). The existence of an adequate regulatory framework, for example, will have a positive effect on entrepreneurship in general (Corrente et al., 2019; Stam & van de Ven, 2019). Likewise, aspects such as legal certainty, intellectual property rights and labour laws, are particularly relevant for KIE (Malerba & McKelvey, 2020). In addition, strong governance can contribute to KIE through financial support as well as the development of university–firm interactions (Corrente et al., 2019; Nicotra et al., 2018). Authors such as Mendonça and Grimpe (2016) and

(a)

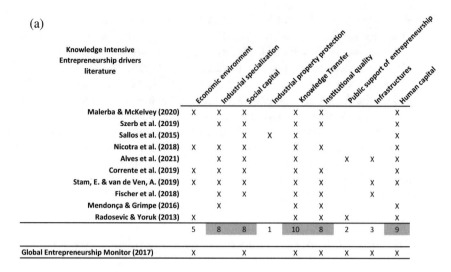

Knowledge Intensive Entrepreneurship drivers literature	Economic environment	Industrial specialization	Social capital	Industrial property protection	Knowledge Transfer	Institutional quality	Public support of entrepreneurship	Infrastructures	Human capital
Malerba & McKelvey (2020)	X	X	X		X	X			X
Szerb et al. (2019)		X	X		X	X			X
Sallos et al. (2015)			X	X	X				X
Nicotra et al. (2018)	X	X	X		X	X			X
Alves et al. (2021)		X	X		X		X	X	X
Corrente et al. (2019)	X	X	X		X	X			X
Stam, E. & van de Ven, A. (2019)	X	X	X		X	X		X	X
Fischer et al. (2018)		X	X		X	X		X	
Mendonça & Grimpe (2016)		X			X	X			X
Radosevic & Yoruk (2013)	X				X	X	X		X
	5	8	8	1	10	8	2	3	9
Global Entrepreneurship Monitor (2017)	X		X		X	X	X	X	X

(b)

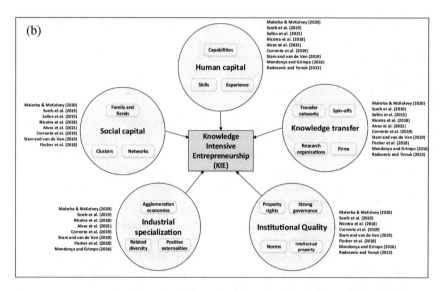

Fig. 2 **a** Drivers behind KIE according to literature on KIE drivers during the past decade. **b** Concept map (see Appendix 1 for more information)

Fischer et al. (2018) point to institutional factors as a possible source of performance differences between countries or regions when it comes to generating entrepreneurial capital. We can therefore establish that institutional quality is an important premise for the proper development of KIE (Nicotra et al., 2018; Szerb et al., 2019).

(ii) **Industrial specialisation.** Because KIE is a systemic phenomenon that is deeply rooted in local contexts (Fischer et al., 2018), a concentration of highly specialised firms and suppliers with similar objectives favours this type of entrepreneurship (Nicotra et al., 2018; Sallos et al., 2016). On the one hand, such specialisation helps the entrepreneur to gain experience and increases his or her capacity to detect market opportunities (Mendonça & Grimpe, 2016; Nicotra et al., 2018). On the other hand, regional innovation systems' organisations and their associated institutions possess complementary knowledge and capabilities that facilitate the emergence of KIE (Malerba & McKelvey, 2020; Martínez-Senra et al., 2013; Szerb et al., 2019). In addition, the clustering of firms generates a critical mass that supports new firms (Fischer et al., 2018; Alves et al., 2021).

(iii) **Social capital.** Because geographical proximity is essential for the transmission of knowledge, the existence of entrepreneur networks and the emergence of clusters facilitate the flow of information, which allows for the efficient distribution of knowledge (Stam & van de Ven, 2019; Alves et al., 2021). First, such knowledge networks enable the sharing of values and resources (Sallos et al., 2016), as well as foster regional learning by reinforcing openness to others' ideas (Szerb et al., 2019). By observing and interacting with others, potential entrepreneurs acquire new skills, gain access to resources and identify forms of financing and customer acquisition strategies (Nicotra et al., 2018; Szerb et al., 2019). Moreover, the existence of channels for sharing existing knowledge increases potential entrepreneurs' potential to perceive new technological opportunities (Malerba & McKelvey, 2020). Second, the presence of industrial groupings that pursue similar objectives (clusters) fosters technological activity at the local level, thus creating infrastructures that are conducive to research and development (R&D) and the transference (Fischer et al., 2018; Sallos et al., 2016).

(iv) **Human capital.** Because KIE is knowledge based, having a skilled pool of highly educated and experienced workers will be key (Alves et al., 2021; Stam & van de Ven, 2019). Authors such as Malerba and McKelvey (2020) find that 72% of entrepreneurs in knowledge-intensive business services (KIBS) have one or more university graduates. The explanation for this may be twofold. On the one hand, the presence of skilled labour and experienced researchers favours the creation of new KIE firms (Mendonça & Grimpe, 2016; Alves et al., 2021). On the other hand, empirical evidence indicates that appropriate training increases the likelihood that potential entrepreneurs will actually take part in entrepreneurial processes, as they are more likely to possess both increased knowledge and differentiating skills. In addition, training increases the potential for perceiving new technological and market opportunities (e.g., León-Mateos et al., 2021; Sartal et al., 2017, 2020).

(v) **Knowledge transfer.** All of the papers identified in the review and the GEM (2017) point to "performance in knowledge spillovers" as an essential factor for KIE entrepreneurship. In fact, although some authors stress the previous drivers to a greater or lesser extent, what they all highlight is that this driver is key for boosting KIE. Experiences such as the Ideon Science Park (Park, 2018), Technopolis Oulu (Nummi, 2007), Silicon Valley, the Boston area and the region around Cambridge in the UK (Lester & Sotarauta, 2007) stress that knowledge transfer played a key role in their success. These experiences not only highlight the importance of knowledge transfer for KIE entrepreneurship but also show the importance of using knowledge transfer networks to transform entrepreneurial initiatives into high-value entrepreneurship based on the relationship between basic research and business (Sallos et al., 2016).

In turn, the interaction between public and private research organisations, as well as between local and regional industries, favours the creation of knowledge-intensive firms (Fischer et al., 2018; Alves et al., 2021). Authors such as Szerb et al. (2019) indicate that the knowledge infrastructure conditions innovative activity, and this activity responds positively to spillovers from university research. In the same vein, KIE is significantly related to academic spin-offs (Alves et al., 2021). Numerous empirical studies conclude that this type of knowledge dissemination is geographically limited (e.g., Keller, 2002), so the presence of local research-oriented universities acts as a fundamental vector for the emergence and development of KIE entrepreneurship. Finally, with regard to non-codified knowledge, the spatial proximity between the owners of knowledge and firms facilitates its dissemination (Szerb et al., 2019). Such cooperation through informal interactions allows much more knowledge to be exchanged than can be contractually specified (Stam & van de Ven, 2019), thus facilitating KIE activity.

From the literature review, it is clear that a consensus exists among authors that a robust institutional environment is key for successful entrepreneurship. Moreover, factors such as well-trained human capital, sectoral specialisation and the establishment of entrepreneur networks and clusters favour the generation of both tacit and explicit knowledge. However, if this knowledge is not transferred to the various actors in the ecosystem (i.e., if no strong and systematic knowledge transfer takes place), KIE entrepreneurship is unlikely to emerge and develop. Thus, based on the SLR and all of the circumstances outlined above, we established our analysis proposition:

- *Knowledge transfer is the key driver for successfully promoting KIE firms, i.e., innovative and creative firms, in a region.*

3 Methods

3.1 Case Study: The Galicia-North Portugal (GNP) Euroregion

To validate our proposition, we needed to carry out a comparative analysis that would allow us to compare the levels of each of the drivers identified in the literature and the KIE performance. For this comparative analysis to be valid, it was essential to identify two regions whose industrial, socioeconomic and institutional characteristics

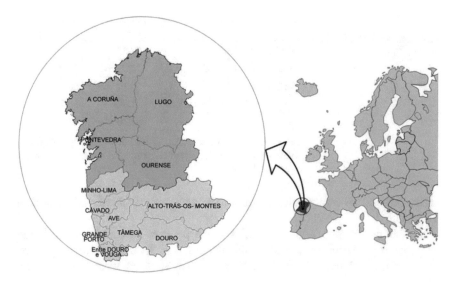

Fig. 3 Map of the Galicia-North Portugal Euroregion

were as similar as possible. After a laborious search, we opted for the Galicia-North Portugal Euroregion (GNP, Fig. 3).

The GNP Euroregion, located in the northwestern part of the Iberian Peninsula (Fig. 3), not only has a common historical and linguistic[1] past but also enjoys strong social, economic and cultural relations today. The territory made up of the two sub-regions together covers a total surface area of 51,000 km^2 (Galicia 29,575, and North Portugal 21,284) and has a population of 6.3 million inhabitants (Galicia 2,702,592 and North Portugal 3,575,338). As researchers such as Vázquez-Gestal et al. (2019, p. 2), among others, have established, "the Galicia-North Portugal Euroregion is not a structure but a concept with an almost psycho-anthropological connotation, in the sense that there are coincidences and ways of sharing vital aspects that do not exist at other borders; language, tradition, culture and history make the territory a continuum, which does not occur with other areas."

Both regions are coastal, having almost 2000 km of continental coastline on the Atlantic Ocean (POOC, 2007), and both feature intense fishing, aquaculture and recreational activity thanks to the numerous rias and estuaries in both areas. In fact, their strategic position in world maritime traffic has given rise to important shared port infrastructures (APVI, 2021). All of these values—plus the activity rate in 2020, which was around 55% in both sub-regions, and plus their gross domestic product

---

[1] Galician-Portuguese was the romance language in the north-western strip of the Iberian Peninsula during the Middle Ages. It gave rise to the Galician (Galicia) and Portuguese (Portugal) spoken today in the GNP Euroregion.

Table 1 Main socioeconomic indicators in the GNP Euroregion (2020)

	Galicia	N. Portugal
Surface area (km^2)	29,571	21,287
Population (no. of inhabitants)	2,702,592	3,575,338
GDP (million €)	59,105.65	60,328.35
GDP per capita (€/inhabitant)	21,870	16,873
Activity rate (%)	52.20%	58.70%

Source Eurostat (2020a, 2020b, 2020c), INEP (2020), IGE (2020). 2020 is used as the base year because it is the last year available for several indicators

(GDP) (around 60 billion euros)—reflect the enormous similarity between them (Table 1).

In addition, from an economic and a social point of view, many cross-border cooperation programmes, including the Interreg VA Spain-Portugal Programme (POCTEP), which has been in force for more than 30 years, have helped to standardise both sub-regions' economies and societies. In fact, these parallels are even reflected in European strategies, where, for example, numerous synergistic plans exist between both regional strategies for addressing cross-border industrial challenges (e.g., RIS3T, 2014).

3.2   Data and Measurements

To carry out our comparative analysis between the two sub-regions that make up the GNP Euroregion, we used various robust official indicators, both for the target indicator (i.e., KIE development) and for the different regional drivers behind KIE's development that have been identified in the literature, as well as control variables (Table 2). In all cases, the statistical classification of the European Union was followed with a NUTS-2[2] level of disaggregation.

First, with respect to the objective indicator, we evaluated the level of KIE performance based on the variation in the number of KIE companies in each year with respect to the previous year in each region. Because our study focused on the industrial domain, we used the KIABI list—the Knowledge Intensive Activities in Business Industries—according to the European Commission's NACE rev.2 report (Appendix 2; Eurostat, 2022a), and which is widely used in the literature (e.g. Smoliński et al., 2015; Sukharev, 2021).

Second, regarding the different KIE antecedents identified in the literature, we used official indicators to measure each of them and to compare their importance in each sub-region (Fig. 3). The measures and constructs used in each case are described below:

---

[2] Nomenclature of territorial units for statistics. The various NUTS levels refer to regions of comparable size within Europe, and the NUTS-2 level is the second-most detailed available.

Table 2 Description of the measures used and the data sources

Variables	Description	Data source	
Target	New KIABI firms	Change in KIABI firms in comparison with the previous year (%)	IGE (2022), INEP (2022), Eurostat (2022a)
Drivers	Industrial specialisation	Employment in technology and knowledge-intensive sectors by NUTS 2 regions and sex (from 2008 onwards, NACE Rev. 2) [HTEC_EMP_REG2_custom_2688953] Percentage of total employment	Eurostat (2022b)
	Social capital	Number of SMEs with innovation cooperation activities with respect to total SMEs (%)—Performance relative to EU in 2014	European Commission (2021)
	Human capital	Number of people (over the total active population) with a third-level educational qualification or jobs involving scientific or technological activities (%)—Performance relative to EU in 2014	European Commission (2021)
	Knowledge transfer	Number of public–private co-authored research publications per inhabitant—Performance relative to EU in 2014	European Commission (2021)
Control variables	Institutional quality	Quality of Government Institute index	EQGI (2022), Charron et al. (2019)
	Creation of new industrial firms	Growth-rate of employment in manufacturing	Eurostat (2022c)

- ***Industrial specialisation*** (%) was measured as the ratio of the level of employment in technology and knowledge-intensive sectors to total employment in the region (Eurostat, 2022b). The higher the ratio, the higher the specialisation in the analysed sector.
- ***Social Capital.*** Following authors such as Müller et al. (2020), we measured this driver using the indicator "innovative SMEs [small and medium-sized enterprises] collaborating with others." This indicator was calculated by dividing the number of SMEs with innovation cooperation activities by the total number of SMEs (European Commission, 2021). It is worth recalling that our focus was on entrepreneurship, so we were particularly interested in assessing the extent to which new enterprises collaborate with other firms. For this reason, we limited this driver to SMEs.
- ***Human capital*** was assessed as the percentage of people with a third-level educational qualification or jobs involving scientific or technological activities, out of the total population aged 15–74 (European Commission, 2021).

- **Knowledge transfer.** Following Hollanders et al. (2019), we assessed this variable using the ratio "number of public–private co-authored research publications per million inhabitants (European Commission, 2021)." This indicator captures public–private research linkages and active collaboration activities between business and public sector researchers that result in academic publications (Mo Ahn et al., 2019).

In addition to these indicators, two measures were used as control measures to rule out their effect on the observed results:

(i) Industrial entrepreneurship was analysed, in global terms. The growth rate of employment for the manufacturing indicator (Eurostat, 2022c) was used as a proxy to obtain this reference value between both sub-regions. The aim was to rule out whether the difference in the creation of new KIE firms was simply a reflection of industrial growth in general or whether it depended on the drivers analysed.

(ii) We used the Quality of Government Institute's index (Charron et al., 2019) to compare institutional quality within the Euroregion. We also used this index to assess whether any differences between the two sub-regions might be behind the different results for the drivers as well as the generation of new KIE firms. This indicator has been widely used for this purpose in several studies (e.g., Khan, 2017; Rodríguez-Pose & Garcilazo, 2015). Moreover, several authors consider that a minimum level of institutional quality is a necessary condition for the emergence of KIE (e.g., Stenholm et al., 2013).

4   Results

Table 3 synthesizes the values collected for each of the sub-regions of the GPN Euroregion during the period of 2015–2019 (latest available data). To facilitate the interpretation of the results, Fig. 4 was constructed in two parts. In the lower part, with the values for Northern Portugal being used as the baseline, we calculate the differential in the four drivers with respect to Galicia.[3] Thus, when the value is higher for Portugal, it appears in the green column—with the value of this difference—and if it is lower than that of Galicia, it is shown in the red columns. In parallel, the upper part of Fig. 4 shows the level of KIE generation (i.e., the variation in the number of industrial KIE companies with respect to the previous year in each region according to Table 2), including the trend line for the period in both cases for a better comparison.

We analyse Fig. 4 in two steps: first the drivers (bottom) and then the KIE evolution (top). With this double analysis, we draw some first conclusions about what may be happening for the best evolution in KIE performance in Portugal.

---

[3] For example, the value of 7.2 in Knowledge Transfer shown in the first green column corresponds to the difference between NP (77.8) and Galicia (70.6). The rest of the columns are calculated in the same way.

Table 3 Data collection for Galicia and North Portugal in the period of 2015–2019

			2015	2016	2017	2018	2019	Period average
Target	Creation of new KIABI firms (%)	Galicia	2.3	4.2	3.1	3.1	3.1	**3.2**
		N. Portugal	4.3	4.3	5.5	3.3	4.7	**4.4**
Drivers	Industrial specialisation (%)	Galicia	2.4	2.7	2.7	2.9	3.1	**2.8**
		N. Portugal	2.1	2	2.5	2.6	2.6	**2.4**
	Social capital (%)	Galicia	90.0	90.0	112.7	112.7	109.6	**103.0**
		N. Portugal	57.8	57.8	73.5	73.5	91.4	**70.8**
	Human capital (%)	Galicia	136.3	136.3	145.5	144.0	153.7	**143.2**
		N. Portugal	73.4	73.4	81.6	84.0	87.9	**80.1**
	Knowledge transfer (publications/Mill. Hab.)	Galicia	70.6	81.3	83.7	84.2	87.3	**81.4**
		N. Portugal	77.8	81.3	85.9	92.4	98.8	**87.2**
Control variables	Creation of new industrial firms (%)	Galicia	− 0.3	5.2	1.5	4.4	2.7	**2.7**
		N. Portugal	3.5	2.8	3.1	2.8	0.9	**2.6**
	Institutional quality index	Galicia	− 0.4	− 0.4	− 0.3	− 0.3	− 0.3	**− 0.4**
		N. Portugal	− 0.1	0.0	0.0	0.0	0.0	**0.0**

The bold values in period average indicates the average value resulting from the time series analyzed

As for the drivers, we see at a glance that the value of three of the drivers (i.e., industrial specialisation, social capital, human capital) always presents a negative balance (red columns) for the NP region. Thus, both industrial specialisation (+ 17% on average, Table 2) and social capital (+ 46%) present significantly higher values in Galicia, even doubling the NP values in the case of human capital (Table 2). Thus, only in one of the drivers, knowledge transfer, does the NP present a better performance for the entire period with an average difference of slightly more than 7% with respect to Galicia (Table 2). It is worth noting that another of the antecedents of KIE development identified in the literature, the institutional quality index, is used here as a control measure supporting the comparability of the two areas (Table 2). The quality index of the Institute of Government places both regions in the same medium to low range (− 0.5 to 0 points) in terms of the quality of both Portuguese and Galician institutions (EQGI, 2022).

Despite these favorable values for Galicia in three of the four drivers, if we go to the upper part of the figure, which analyses the value of the creation of new KIE companies, we observe that the results are significantly higher in northern Portugal. Thus, the creation of KIE companies in Portugal (4.4%) is 37.5% higher than in

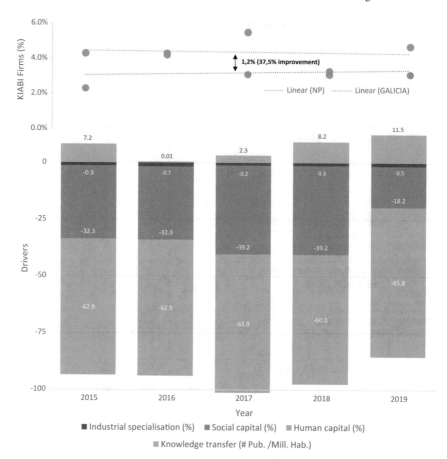

Fig. 4 Comparison of KIE evolution (upper) and drivers (bottom) in the GNP Euroregion (2015–2019)

Galicia (3.2%), where, moreover, the figure shows a certain negative trend in recent years. In fact, if we quantitatively analyse these values for both countries, this means the creation of 3246 new KIE firms in the industrial field in Galicia, whereas in Portugal, they have been almost triple (8479) for the same period (Eurostat, 2022a). The importance of this difference is even more relevant if we take into account that the creation of industrial companies (control measure) is higher in Galicia (2.7%) compared with the NP (2.6%).

These results support our research proposal: under similar conditions of institutional quality, regional knowledge transfer could be the key driver for successfully promoting KIE firms in any region. In Fig. 3, we observe that despite the fact that the NP presents worse performance for the other three drivers throughout the entire period, standing out in the transfer section could be the determining factor for achieving the best KIE performance. Therefore, in line with the literature, our exploratory analysis seems to show that if knowledge transfer in a region does not show relevant performance, the other drivers might not have relevant effects on the creation of KIE firms and, consequently, on that region's economic development.

5 Discussion and Conclusions

Our work is aimed assessing the importance of the main key drivers behind KIE firms' development by means of a case study in the GNP Euroregion. The comparative analysis proposed seems to show that under similar socio-economical and institutional conditions for the main drivers of this type of innovative and creative firms, a higher level of regional knowledge transfer will have a decisive impact. Indeed, because KIE firms are widely recognised in the literature as the type of companies that have the greatest impact on regions' economic and social development (Malerba & McKelvey, 2020; Szerb et al., 2019), our findings allow us to establish a positive relationship between knowledge transfer in a region and its greater economic development.

These findings not only add empirical evidence for each KIE driver to the existing literature but also may be useful for designing more efficient public policies to promote regional development in the field of Industry 5.0. Although the other drivers (e.g., industrial specialisation, human capital, etc.) are also important and should be promoted, our findings highlight the need to pay special attention to the knowledge transfer driver. In this new context, known as 5.0, the public administration must foster the transfer of knowledge among the innovation system's various agents to boost the regions' long-term development.

It must favour knowledge spillovers from companies and research agents, the systematising of this process, and the establishment of collaboration links that potential entrepreneurs can benefit from. Likewise, public aid should promote the inclusion of scientists in entrepreneurship projects, as this could help to intensify such knowledge transfer. Several Portuguese regional initiatives (e.g., Incentives System for Research and Technical Development, or the PROCIÊNCIA 2020 programme) could serve as examples for promoting greater cooperation and knowledge exchange between the academic and business worlds in other Euroregions.

Appendix 1

KIE drivers according to literature during the past decade (in detail).

	Economic environment	Spatial externalities and industrial specialization	Clustering, networking and social capital	Industrial property protection	Knowledge spillovers, universities and innovation	Institutional quality	Public support of entrepreneurship	Infrastructures	Education, human capital and creativity
Malerba and McKelvey (2020)	Access to finance: Bank credit, venture capital	Regional innovation system, specific industrial clusters	Specific sectoral knowledge actors and institutions, specific industrial clusters, knowledge networks		Knowledge networks. Universities and research organizations play a crucial role in creating and transferring knowledge	Entrepreneurs are highly dependent upon the institutional context			Education facilitates knowledge acquisition
Szerb et al. (2015)		Spatial externalities (agglomeration economies, population growth, size of the potential regional market)	Clustering, networking, social capital		Knowledge spillovers, universities and innovation	The role of the state: (Quality of government, regulation, corruption)			Education, human capital and creativity
Sallos et al. (2016)			Social capital effect	Mechanisms to protect innovations: patenting, secrecy, trademarks…	Internal and external knowledge sources: customers, suppliers, competitors				Human resources

(continued)

(continued)

	Economic environment	Spatial externalities and industrial specialization	Clustering, networking and social capital	Industrial property protection	Knowledge spillovers, universities and innovation	Institutional quality	Public support of entrepreneurship	Infrastructures	Education, human capital and creativity
Nicotra et al. (2018)	Financial capital: Accesible markets, funding and finance	Entrepreneurial ecosystem	Social capital: networking, cultural support		Knowledge capital, university research	Institutional capital: Policy regulation and norms, support structure			Qualified human capital
Alves et al. (2021)		Business and market dynamics: business concentration, distance to economic hub, positive externalities of agglomeration	Cultural and social norms		Science and technology: university research, patents, knowledge transfer		Public and private capital flows, multinational investment	Infrastructure. Support system	Human development and education. Human capital
Corrente et al. (2019)	Financial support	Entrepreneurial ecosystem	Cultural and social norms		R&D transfer	National policy			Education
Stam and van de Ven (2019)	Financial resources	Entrepreneurial ecosystem, spatial context, industrial infrastructure	Social networks		Knowledge	Formal institutions, informal institutions		Physical infrastructure	Culture, Talent, Leadership

(continued)

(continued)

	Economic environment	Spatial externalities and industrial specialization	Clustering, networking and social capital	Industrial property protection	Knowledge spillovers, universities and innovation	Institutional quality	Public support of entrepreneurship	Infrastructures	Education, human capital and creativity
Fischer et al. (2018)		Agglomeration dynamics	Economic hubs		Knowledge & innovation system	Local market conditions		Access to economic hubs	
Mendonça and Grimpe (2016)		Specialization and diversity of the regional skill base may benefit entrepreneurship			Knowledge spillovers	Institutional factors favour the emergence of entrepreneurial capital			The generation and absorption of new knowledge requires qualified human capital,
Radosevic and Yoruk (2013)	Financing of innovation projects				Knowledge development and dissemination. Knowledge networks	Institutional opportunities: creating an changing insitutions (e.g. IPR laws, tax laws, environment and safety regulations	Market opportunities: financing for innovation projects, R&D subsidies, market for knowledge based services		Provision of education and training, creation of human capital, production and reproduction of skills
	5	8	8	1	10	8	2	3	9

(continued)

(continued)

	Economic environment	Spatial externalities and industrial specialization	Clustering, networking and social capital	Industrial property protection	Knowledge spillovers, universities and innovation	Institutional quality	Public support of entrepreneurship	Infrastructures	Education, human capital and creativity
Global Entrepreneurship Monitor (2017)	Financial support. internal market dynamics, internal market openness		Cultural and social norms		R&D transfer	National policy, regulation, commercial infrastructure	Government programmes	Physical Infrastructure	Education, higher education

Appendix 2

KIABI list. Sectors and sub-sectors included in KIE activity for industrial field (Eurostat, 2022a).

NACE Rev. 2 Codes	Description
09	Mining support service activities
19	Manufacture of coke and refined petroleum products
21	Manufacture of basic pharmaceutical products and pharmaceutical preparations
26	Manufacture of computer, electronic and optical products
51	Air transport
58	Publishing activities
59	Motion picture, video and television programme production and pharmaceutical preparations
60	Programming and broadcasting activities
61	Telecommunications
62	Computer programming, consultancy and related activities
63	Information service activities
64	Financial service activities, except insurance and pension funding
65	Insurance, reinsurance and pension funding, except compulsory social security
66	Activities auxiliary to financial services and insurance activities
69	Legal and accounting activities
70	Activities of head offices; management consultancy activities
71	Architectural and engineering activities; technical testing and analysis
72	Scientific research and development
73	Advertising and market research
74	Other professional, scientific and technical activities
75	Veterinary activities
78	Employment activities
79	Travel agency, tour operator reservation service and related activities
90	Creative, arts and entertainment activities

References

Achim, M. V., Borlea, S. N., & Văidean, V. L. (2021). Culture, entrepreneurship and economic development. An empirical approach. *Entrepreneurship Research Journal, 11*(1).

Alves, A. C., Fischer, B. B., & Vonortas, N. S. (2021). Ecosystems of entrepreneurship: Configurations and critical dimensions. *The Annals of Regional Science*, 1–34.

Aparicio, S., Audretsch, D., & Urbano, D. (2021). Does entrepreneurship matter for inclusive growth? The role of social progress orientation. *Entrepreneurship Research Journal, 11*(4).

APVI. (2021). Autoridad Portuaria de Vigo. Retrieved from: https://www.apvigo.es/

Australian Government. (2021). Entrepreneurs' programme. Retrieved from: https://business.gov. au/grants-and-programs/entrepreneurs-programme

Caloghirou, Y., & Llerena, P. (2015). Public policy for knowledge intensive entrepreneurship: Implications from the perspective of innovation systems. In *Dynamics of knowledge intensive entrepreneurship* (pp. 451–463). Routledge.

Charron, N., Lapuente, V., & Annoni, P. (2019). Measuring quality of government in EU regions across space and time. *Papers in Regional Science.* https://doi.org/10.1111/pirs.12437

Corrente, S., Greco, S., Nicotra, M., Romano, M., & Schillaci, C. E. (2019). Evaluating and comparing entrepreneurial ecosystems using SMAA and SMAA-S. *The Journal of Technology Transfer, 44*(2), 485–519.

EQGI. (2022). European quality of government index. University of Goteborg. Retrieved from: https://www.gu.se/en/quality-government/qog-data/data-downloads/european-quality-of-government-index

European Commission. (2021). European and regional innovation scoreboard 2021. Retrieved from: https://ec.europa.eu/research-and-innovation/en/statistics/performance-indicators/european-innovation-scoreboard/eis

Eurostat. (2020a). Regional statistics by NUTs classification. Population on 1 January by NUTS 2 region. Retrieved from: https://ec.europa.eu/eurostat/databrowser/view/TGS00096__custom_2492917/default/table?lang=en

Eurostat. (2020b). Regional statistics by NUTs classification. Gross domestic product (GDP) at current market prices by NUTS 2 regions. Retrieved from: https://ec.europa.eu/eurostat/databrowser/view/nama_10r_2gdp/default/table?lang=en

Eurostat. (2020c). General and regional statistics; Regional statistics by NUTS classification; Area by NUTS 3 region. Retrieved from: https://ec.europa.eu/eurostat/databrowser/view/reg_area3/default/table?lang=en

Eurostat. (2022a). Eurostat indicators on high-tech industry and knowledge—Intensive services. Annex 8—Knowledge Intensive Activities by NACE Rev. 2. Retrieved from https://ec.europa.eu/eurostat/cache/metadata/Annexes/htec_esms_an8.pdf

Eurostat. (2022b). General and regional statistics; Regional statistics by NUTS classification; Regional science and technology statistics. Employment in high technology. Retrieved from https://ec.europa.eu/eurostat/databrowser/view/HTEC_EMP_REG2__custom_2688953/default/table?lang=en

Eurostat. (2022c). General and regional statistics; Regional statistics by NUTS classification; Regional structural business statistics; SBS data by NUTS2 regions and NACE rev.2. Retrieved from https://ec.europa.eu/eurostat/databrowser/view/SBS_R_NUTS06_R2__custom_2763644/default/table?lang=en

Fischer, B. B., Queiroz, S., & Vonortas, N. (2018). On the location of knowledge-intensive entrepreneurship in developing countries: Lessons from São Paulo, Brazil. *Entrepreneurship & Regional Development, 30*(5–6), 612–638.

Foss, N. J., & Saebi, T. (2017). Fifteen years of research on business model innovation: How far have we come, and where should we go? *Journal of Management, 43*(1), 200–227.

Gao, L., Melero, I., & Sese, F. J. (2019). Multichannel integration along the customer journey: A systematic review and research agenda. *The Service Industries Journal*, 1–32.

GEM. (2017a). Global Report 2016/17. http://www.gemconsortium.org/report/49812

GEM. Global Entrepreneurship Monitor. (2017b). Global report 2016/17. Retrieved October 12, 2020, from http://www.gemconsortium.org/report/4981

Government of Canada. (2021). Innovative Solutions Canada. Recuperado el 05-08-2021 de https://www.ic.gc.ca/eic/site/101.nsf/eng/home

Hollanders, H., Es-Sadki, N., & Merkelbach, I. (2019). Regional innovation scoreboard methodology report. Retrieved from https://ec.europa.eu/docsroom/documents/37783

IGE. (2020). Instituto Galego de Estatística. Taxa de actividade por nacionalidade, sexo e nivel de formación alcanzado (CNED-2014). Retrieved from: https://www.ige.eu/igebdt/selector.jsp?COD=6531&AT=1

IGE. (2022). Instituto Galego de Estadística. Economía; Directorio de empresas e unidades locais; Altas, baixas e permanencias por actividade (grupos CNAE 2009). Retrieved October 27, 2020, from https://www.ige.eu/igebdt/selector.jsp?COD=8475&paxina=001&c=0307006001

INEP. (2020). Instituto Nacional de Estatística (Portugal). Retrieved from https://www.ine.pt/xportal/xmain?xpid=INE&xpgid=ine_indicadores&indOcorrCod=0006175&xlang=pt&contexto=bd&selTab=tab2

INEP. (2022). Instituto Nacional de Estatística (Portugal). Empresas por localizaçao geográfica e atividade económica. Retrieved from: https://www.ine.pt/xportal/xmain?xpid=INE&xpgid=ine_indicadores&indOcorrCod=0008466&contexto=bd&selTab=tab2

Kasabov, E. (2016). When an initiative promises more than it delivers: A multi-actor perspective of rural entrepreneurship difficulties and failure in Thailand. *Entrepreneurship & Regional Development, 28*(9–10), 681–703.

Keller, W. (2002, March). Trade and the transmission of technology. *Journal of Economic Growth, 7*(1), 5–24.

Khan, H. A. (2017). Globalization and the quality of government: An analysis of the relationship. *Public Organization Review, 17*, 509–524. https://doi.org/10.1007/s11115-016-0352-4

Kyllingstad, N. (2021). Overcoming barriers to new regional industrial path development: The role of a centre for research-based innovation. *Growth and Change, 52*(3), 1312–1329.

Lassen, A. H., McKelvey, M., & Ljungberg, D. (2018). Knowledge-intensive entrepreneurship in manufacturing and creative industries: Same, same, but different. *Creativity and Innovation Management, 27*(3), 284–294.

León-Mateos, F., Sartal, A., López-Manuel, L., & Quintas, M. A. (2021). Adapting our sea ports to the challenges of climate change: Development and validation of a port resilience index. *Marine Policy, 130*, 104573.

Lester, R., & Sotarauta, M. (2007). Innovation, universities and the competitiveness of regions. *Technology Review, 214*, 2007.

Malerba, F., & McKelvey, M. (2020). Knowledge-intensive innovative entrepreneurship integrating Schumpeter, evolutionary economics, and innovation systems. *Small Business Economics, 54*(2), 503–522.

Martínez-Senra, A. I., Quintás, M. A., Sartal, A., & Vázquez, X. H. (2013). ¿ Es rentable «pensar por pensar»? Evidencia sobre innovación en España. *Cuadernos De Economía y Dirección De La Empresa, 16*(2), 142–153.

Mendonça, J., & Grimpe, C. (2016). Skills and regional entrepreneurship capital formation: A comparison between Germany and Portugal. *The Journal of Technology Transfer, 41*(6), 1440–1456.

Mo Ahn, J., Roijakkers, N., Fini, R., & Mortara, L. (2019). Leveraging open innovation to improve society: Past achievements and future trajectories. *R&D Management, 49*(3).

Müller, J. M., Buliga, O., & Voigt, K. I. (2020). The role of absorptive capacity and innovation strategy in the design of industry 4.0 business Models—A comparison between SMEs and large enterprises. *European Management Journal, 49*(103921). https://doi.org/10.1016/j.emj.2020.01.002

Nicotra, M., Romano, M., Del Giudice, M., & Schillaci, C. E. (2018). The causal relation between entrepreneurial ecosystem and productive entrepreneurship: A measurement framework. *The Journal of Technology Transfer, 43*(3), 640–673.

Nightingale, P., & Coad, A. (2014). Muppets and gazelles: Political and methodological biases in entrepreneurship research. *Industrial and Corporate Change, 23*(1), 113–143.

Nummi, J. (2007). University–industry collaboration in medical devices development: Case study of the Oulu Region in Finland. *Innovation, Universities, and the Competitiveness of Regions, 214*(2007), 95.

Park, S.-C. (2018). The fourth industrial revolution and implications for innovative cluster policies. *AI & SOCIETY, 33*, 433–445. https://doi.org/10.1007/s00146-017-0777-5

POOC. (2007). Plano de Ordenamento da Orla Costeira Caminha–Espinho (POOC). Agência Portuguesa do Ambiente. Retrieved from: https://apambiente.pt/agua/pooc-caminha-espinho

Pugh, R., MacKenzie, N. G., & Jones-Evans, D. (2018). From 'Techniums' to 'emptiums': The failure of a flagship innovation policy in Wales. *Regional Studies, 52*(7), 1009–1020.

Radosevic, S., & Yoruk, E. (2013). Entrepreneurial propensity of innovation systems: Theory, methodology and evidence. *Research Policy, 42*(5), 1015–1038.

RIS3T. (2014). Cross-border smart specialisation strategy of Galicia-Northern Portugal. Galician Innovation Agency (GAIN) and the Regional Coordination and Development Commission of Northern Portugal (CCDRN). Retrieved from http://www.ris3galicia.es/wp-content/uploads/2016/07/RIS3T_INGLES.pdf

Rodríguez-Pose, A., & Garcilazo, E. (2015). Quality of government and the returns of investment: Examining the impact of cohesion expenditure in European Regions. *Regional Studies, 49*(8), 1274–1290. https://doi.org/10.1080/00343404.2015.1007933

Sallos, M., Yoruk, E., & Garcia-Perez, A. (2016). A business process improvement framework for knowledge-intensive entrepreneurial ventures. *Journal of Technology Transfer, 42*(2), 354–373. https://doi.org/10.1007/s10961-016-9534-z

Sartal, A., Martínez-Senra, A. I., & García, J. M. (2017). Balancing offshoring and agility in the apparel industry: Lessons from benetton and inditex. *Fibres & Textiles in Eastern Europe.*

Sartal, A., Carou, D., & Davim, J. P. (Eds.). (2020). *Enabling technologies for the successful deployment of Industry 4.0.* CRC Press.

Schumpeter, J. (1934). *The theory of economic development.* Harvard Economic Studies. ISBN: 9780674879904.

Sivarajah, U., Kamal, M. M., Irani, Z., & Weerakkody, V. (2017). Critical analysis of Big Data challenges and analytical methods. *Journal of Business Research, 70*, 263–286.

Smoliński, A., Bondaruk, J., Pichlak, M., Trząski, L., & Uszok, E. (2015). Science-economy-technology concordance matrix for development and implementation of regional smart specializations in the Silesian Voivodeship, Poland. *The Scientific World Journal, 2015.*

Stam, E., & van de Ven, A. (2019). Entrepreneurial ecosystem elements. *Small Business Economics*, 1–24.

Stenholm, P., Acs, Z. J., & Wuebker, R. (2013). Exploring country-level institutional arrangements on the rate and type of entrepreneurial activity. *Journal of Business Venturing, 28*(1), 176–193.

Sukharev, O. (2021). Measuring the contribution of the "knowledge economy" to the economic growth rate: Comparative analysis. *Journal of the Knowledge Economy, 12*(4), 1809–1829.

Szerb, L., Lafuente, E., Horváth, K., & Páger, B. (2019). The relevance of quantity and quality entrepreneurship for regional performance: The moderating role of the entrepreneurial ecosystem. *Regional Studies, 53*(9), 1308–1320.

Szerb, L., Ács, Z. J., Komlósi, É., & Ortega-Argilés, R. (2015). Measuring entrepreneurial ecosystems: The regional entrepreneurship and development index (redi). DiscussionPaper# CFE-2015-02]. Henley Centre for Entrepreneurship.

Tranfield, D., Denyer, D., & Smart, P. (2003). Towards a methodology for developing evidence-informed management knowledge by means of systematic review. *British Journal of Management, 14*(3), 207–222.

U.S. Economic Development Administration's (EDA). (2021). Build to scale program. Retrieved from https://eda.gov/oie/buildtoscale/

Urbano, D., & Aparicio, S. (2016). Entrepreneurship capital types and economic growth: International evidence. *Technological Forecasting and Social Change, 102*, 34–44.

Vázquez-Gestal, M., Fernández-Souto, A. B., & Puentes-Rivera, I. (2019). The Galicia-North Portugal Euroregion in its Universities problems and inconveniences for visibility. In *2019 14th Iberian Conference on Information Systems and Technologies (CISTI)* (pp. 1–6). IEEE.

Veenhoven, R., & Vergunst, F. (2014). The Easterlin illusion: Economic growth does go with greater happiness. *International Journal of Happiness and Development, 1*(4), 311–343.

Wynn, M., & Jones, P. (2019). Context and entrepreneurship in knowledge transfer partnerships with small business enterprises. *The International Journal of Entrepreneurship and Innovation, 20*(1), 8–20.

A Brief Glance About Recruitment and Selection in the Digital Age

Nara Caroline Santos Silva and Carolina Feliciana Machado🄓

Abstract Recruitment and selection are one of the most performed practices in companies by the human resources sector. Currently, organizations are increasingly reinventing themselves and updating their methods with the advancement of technology. This chapter aimed to investigate the impact of the digital age on the recruitment and selection process within organizations. Through the literature review, we sought to point out the advantages and disadvantages of the online recruitment and selection process through digital platforms and social networks. The results allowed us to conclude that organizations are employing new technologies and using social networks to recruit people, and that this occurrence has everything to expand in the coming years. It has been proven that the recruitment and selection procedures on the internet enable time and cost savings, however the excess of information presents a lack of reliability for recruiters. Social networks are seen by companies as a complement to information and not as a substitute for other traditional recruitment and selection methods.

Keywords Recruitment · Selection · Online recruitment · Social networks

1 Introduction

In the current period of global networking, information technology has completely inspired human resource management processes and human resource management departments. For over a decade, digital possibilities have challenged traditional ways of providing human resource management services within businesses and organizations.

N. C. S. Silva · C. F. Machado (✉)
School of Economics and Management, University of Minho, Braga, Portugal
e-mail: carolina@eeg.uminho.pt

C. F. Machado
Interdisciplinary Centre of Social Sciences (CICS.NOVA.UMinho), University of Minho, Braga, Portugal

The recruitment and selection processes through the internet are one of the most adopted in the current scenario, having as its main pillar digital platforms such as Facebook, LinkedIn, among others, which help in the search for the best candidates. One of the main characteristics of this style of process is its feasibility, since the technological reality is part of the daily life of the world population, facilitating the process of those who hire and those who want to be hired.

According to Costa (2012: 38) "online capture and selection is understood as any and all tools or systems that use the Web or the Internet to collect information about candidates, with the purpose of helping in hiring decisions". In this way, recruitment through digital tools will continue to grow as a greater number of people have access to the online tools and are compatible to enter the job market.

In this bias, this work was carried out with the intention of studying and, concretely, understanding the impact of the digital age on the recruitment and selection process within organizations.

Thus, the main aim of this chapter is to investigate the relevance of using online tools in the recruitment and selection processes for the human resources area. For this, a brief theoretical explanation was made of both methods in the traditional model, as well as online. The research was a source of primary interest in the investigation of the advantages and disadvantages of the organizational recruitment and selection processes in the digital environment.

2   Recruitment

Recruitment is defined as the process of attracting a set of candidates for a specific position. It consists of the search and attraction of people for a certain function (Chiavenato, 2005). It is not just a process to hire the best among those who have applied for a certain position, but rather a matter of enrolling the right candidate in human capital (Dhamija, 2012).

For Marras (2007), recruitment takes place within the context of the labor market, recognized as the space where exchanges take place between organizations and people who are willing to work and also those who are already active and introduced in the labor market.

Milkovich and Boudreau (1994) state that recruiting is the way to identify and obtain a pool of candidates, some of whom will be selected to receive job offers.

They still believe that recruiting is not just important to the organization, but a two-way communication resource. Because it is at this time that candidates want accurate information about what it would be like to work in a particular organization. And on the other hand, organizations want to know precisely what kind of employees' candidates would be if they were contacted.

In general, both job applicants and organizations send signals about the employment relationship. Candidates signal that they are attractive and suitable to fit the vacancy offered and organizations indicate that they are good places to work, so they aim to receive signals from candidates that are issuing true information.

One of the most important issues of the recruitment phase is the identification, selection and maintenance of specific recruitment sources for the organization, which serve to improve the recruitment process, by increasing the number of suitable candidates, reducing recruitment time and decreasing recruitment costs, thus saving for the application of techniques (Chiavenato, 2005).

The aforementioned author also emphasizes that the fundamental sources of recruitment for organizations are carried out through internal and external research. Internal research is focused on the human resources needs of organizations and the respective policies to be adopted in relation to employees, otherwise external research aims to research the human resources market in order to support the function of the organizations interests at the candidate levels that they want to attract.

For Breaugh and Starke (2000), recruitment affects both the number and the profiles of people who are determined to apply for, or obtain, a job opportunity. Recruitment becomes strategic when people management professionals answer five essential questions: who to recruit, where to recruit, what resources to use, when to recruit, and what message to communicate.

Recruitment can be classified as: internal, external or mixed. The intern privileges the companies' own employees, is fast and provides low costs; the external one takes place with candidates who are not part of the organization and who can bring new ideas to the work environment; and finally, mixed recruitment encompasses internal and external hiring (Coradini & Murini, 2009).

Research shows that over the years' researchers have turned their attention to the effectiveness of various recruitment methods. For example, comparisons of individuals who have been referred by a current employee and who have applied themselves to those who are referred by an employment agency. They reached several conclusions, however two of them drew more attention from the authors (Zottoli & Wanous, 2000).

These authors emphasize that the realistic hypothesis suggests that individuals recruited through certain methods, such as employee referrals, have a more precise understanding of what employment position involves them. The individual difference hypothesis (those who are referred by agency) posits that different recruitment methods can bring a job by opening attention to different types of individuals who vary in important attributes (e.g. skill, work ethic).

Recruitment is enabled to: produce labor cost capacity and/or increase customer perception of the company's products and services; identify and search for talents that are rare in the job market; contribute to the fact that recruitment processes are not easy to copy; establish an innovative and characteristic recruitment strategy for the organization, making it irreplaceable; match other Human Resource Management practices, such as recognition, selection and development (Orlitzky, 2007; cited in Cassiano et al., 2016).

Recent recruitment practices show that job seekers, as well as professional recruiters and organizations that need staff, are increasingly using the internet as a resource in the selection process (Furtmuellera et al., 2011). These authors add that recruiters are using internet postings and surveys to advertise work, while job seekers are handling to apply online.

Online recruitment allows access to the extensive amount of information of potential candidates at a reduced cost when compared to disclosures in traditional newspapers and magazines (Lievens & Chapman, 2009; cited in Cassiano et al., 2016).

The great advantage of online recruitment is the possibility for candidates to publish and disseminate information about their professional path, experiences and career goals in a durable way, 24 h a day and globally. Facebook, the most popular network on the planet, with more than 750 million users worldwide, offers a variety of solutions in the area of recruitment (Lusa, 2015).

According to Rajani (2016), the costs and time in hiring a person online are lower than traditional methods. The same author cites that in a study carried out by Recruitsoft/iLogos, research shows among five hundred companies that when a company uses a traditional recruitment cycle, it takes an average of 43 days. However, they could reduce it to six days by posting job advertisements online rather than using newspapers.

3 Selection

Following recruitment, selection takes place later, in a process of adding people to the organization. Characterized by a typology that will enable the appropriate choice of the profile necessary for the required function, that is, it is a set of technical actions that aim to meet the needs of an organization's professionals (Coradini & Murini, 2009).

Personnel selection can be defined simply as choosing the right person for the right position, or among recruited candidates, those who are the most suitable for the existing positions in the organization, aiming to maintain or increase the efficiency and performance of the personnel, as well as the effectiveness of the organization. Thus, allowing to solve two basic problems: adaptation of the person to the position and the organization and vice versa and the efficiency and satisfaction of the person in the function (Chiavenato, 1999).

To Martins (2007: no page), "selection consists, first of all, in the comparison between candidates' profiles and the requirements of the position or function; the ideal is that the profile and the function fit. Thus, it is necessary to choose the right person for the right position, that is, among recruited candidates those most suitable for the existing positions in the company, in order to maintain or increase the efficiency and performance of the staff."

Each selection phase represents a moment of choice, aiming to increase the organization's knowledge of the employee's experiences, skills and motivation, thus expanding the information so that the final selection can be made. All procedures used for the selection are valid and adequate as long as they are handled by trained and responsible professionals, considering the objectives of each one.

Usually, the selected candidates are the ones that best satisfy the needs of the organization and that, at the same time, are indicators of a favorable performance in the position to be filled. In this way, the identified characteristics of the candidates are important for the use of several selection techniques together and for analyzing their results, before taking an effective decision regarding the most suitable individual for the role (Cunha et al., 2010).

In the personnel selection process, the indispensable objective is to choose a candidate who has superior knowledge and skills to perform the activities of a certain position, as well as to help the organization achieve its objective, given the demands of the job market (Katsurayama et al., 2012).

In short, the selection process seeks to reach a conclusion of analysis of skills, attitudes, knowledge, personality and some other elements that are linked to the adjustment more specific to the organization such as: sex, physical size, height, address, ownership of a car and age (Coradini & Murini, 2009).

According to Guimarães and Arieira (2005), the most used selection methods are group dynamics and interviews. Group dynamics takes place in groups and allows the evaluator to observe the candidate's behavior and how he relates and interacts with others in the group. The formal interview, on the other hand, allows employers/interviewers to find out who the best qualified candidates are through an individualized assessment.

Regularly, using only curriculum analysis is unsatisfactory to have a complete view of the candidate, as many candidates fail to make a good impression through their CVs. They end up not putting essential information, such as: English proficiency. Therefore, it is advisable to associate its use with other techniques, such as the interview (Coradini & Murini, 2009).

4 Online Recruitment and Selection

As pointed out by Lima and Rabelo (2018), nowadays both companies and professionals in the field of recruitment and selection tend to show greater concern with the technique that they will develop in their selection procedures than, really, with the collaboration that it can offer to improve the nature of work and the company-candidate relationship.

Given the easy access to information and instant communication, it can be said that people management has undergone a wide change after handling the media or social networks. Employees considered to be from the digital age (from 1995 onwards) are unaware of a world without the internet and all its communicational benefits (Milreu, 2009; cited in Cembranel et al., 2013).

Sharing the idea of the authors mentioned above, the advancement of technology and the dissemination of the internet in all areas of communication began to be noticed that new technologies, namely social networks, enable an increase in the number of candidates compared to other traditional methods. The internet makes it possible to effectively find a never-ending source of talent, reaching a wider audience of job seekers.

In terms of human resources management, the internet has transformed the way of recruiting from both perspectives, that is, both organizations and candidates for the job on offer. One of the most popular non-traditional forms of recruitment is e-recruitment (Dhamija, 2012).

One of the first doubts that arose at the beginning of the digitization of the recruitment and selection processes is whether the use of digital platforms would be really effective. Given this scenario, researchers investigated the connection between the type of employment of employees and their work performance (Suvankulov, 2013).

The aforementioned author found that this relationship is positive when recruitment takes place online, as people acquire more detailed information about the environment and work tasks of the organizations for which they apply on the Internet. Consequently, the fit between candidate and company is greater.

For Vieira (2010) companies will be able to use the internet to provide people with specific information about their company, its method of how it works, among other possible information that approves whether the company will be promoted and recognized by potential candidates.

Recruiters are faced day after day with different challenges while working in the digital environment (online) and compared to printing (offline) of the CVs. Digital CVs can be used in processes such as automated storage, research, pre-selection, comparison and ranking of candidates. In addition to being easily transferred to other systems, it is relatively easy to update their contents. As well, digital CVs offer the opportunity to send instant responses to candidates (Furtmuellera et al., 2011).

Adduces Vieira (2010) that a well-executed recruitment strategy well established on the Internet and other recruitment methods will attract excellent candidates, simplifying the time and cost of hiring, with effectiveness in the adequacy of responses and a significant improvement in hiring decisions. hiring, as the two types of methods provide very important information that together provide a correct recruitment.

Recruiters are faced day after day with different challenges while working in the digital environment (online) and compared to printing (offline) of the resumes. Digital CVs can be used in processes such as automated storage, research, pre-selection, comparison and ranking of candidates. In addition to being easily transferred to other systems, it is relatively easy to update their contents. As well, digital resumes offer the opportunity to send instant responses to candidates (Furtmuellera et al., 2011).

Vieira (2010) states that a well-executed recruitment strategy, well established on the Internet and other recruitment methods, will attract excellent candidates, simplifying the time and cost of hiring, with effectiveness in the adequacy of responses and a significant improvement in hiring decisions, as the two types of methods provide very important information that together provide a correct recruitment.

The terms e-recruitment, online recruitment, cyberecruiting or internet recruitment are similar expressions. They imply an explicit distribution of jobs online, being a complete process that includes job advertisements, receiving CVs and building a human resources database with candidates and holders (Dhamija, 2012).

According to Cassiano et al. (2016) four tools stand out in the practice of online recruitment, these are recognized as: company websites, career portals, job boards and, more recently, social networks. The employment websites give the possibility of publicizing organizational positions to a wider audience, with low cost and wide access to CVs in a database.

The practice of e-recruitment and online selection can be met in various demands that are currently presented to the area of human resources. In addition to contributing to a positive effect, mainly, the relationship of approximation and easy communication between candidates and companies, such as "greater dynamics and optimization of time in processes and results" (Lima & Rabelo, 2018: 147).

Dhamija (2012) argues that e-recruitment has become a significant part of the recruitment strategy. It can be used to track and manage applicant applications, especially among larger organizations. It can also provide some remarkable benefits in terms of efficiency and cost.

5 Recruitment and Selection on Social Networks

It was found that social networks are efficient in the search and identification of candidates in the recruitment processes in the perception of company recruiters (Cassiano et al., 2016).

Studies show that LinkedIn and Facebook, although used differently, are used by decision makers, both in small and medium-sized companies and in large organizations during the recruitment and selection processes (Caers & Castelyns, 2011).

Regarding the advantages of using social networks in recruitment, the time spent, the costs and the efficiency of the platforms are considered to be the main added value in the recruitment process. It also shows that they are currently used as an important recruitment tool and that, combined with traditional methods, they can generate valuable results for organizations (Freitas, 2017).

In the recruitment phase, LinkedIn is more used than Facebook, as the latter is considered less professional than the former, to communicate vacancies to the outside world and to actively search for candidates (Caers & Castelyns, 2011).

For these authors, during the selection phase, LinkedIn and Facebook are used by many interviewees to increase the volume of information available for the selection interview and for a minority of decision makers (in this case, those who are selecting) to decide on the invitations to a first selection interview.

The authors also note the importance of greater awareness among candidates of the effects that their social media accounts can have on the success of their application, both in a positive and negative sense. Recruiters will be aware if what is posted on the networks matches the candidate's real profile. In addition, the formulation of organizational policies must therefore be able to contribute to the success of recruitment and selection efforts in the age of the social network.

6 Advantages and Disadvantages of Online Recruitment and Selection

Online recruitment brings advantages not only for the company, but also for those recruited (Costa, 2018). The online recruitment process is beneficial both for the candidate, who has their information available in real time and globally, and for the company, which accelerates the entire process of recruiting and selecting, which proves to be one of the biggest advantages of the process of online R&S, which is the reduction of time (Godinho, 2009).

In agreement with the same idea Kim and O'Connor (2009) point out that the Internet intensifies the speed of the recruitment process, enabling those responsible for recruitment to post ads 24 h a day and in more than one recruitment source at a time. Likewise, potential candidates can send their CVs 24 h a day, from anywhere.

Costa (2018) found in his study that the advantages of online recruitment are related to time, cost, search breadth and security variables. As disadvantages, the impersonality of the process is notable, which Costa (2012) points out as a problem regarding the lack of direct contact with the candidate, which can cause a lack of reliability in the process.

Another factor that companies do not consider reliable is that the information available on candidates' networks is often not very thorough (Vieira, 2010).

Costa (2018) emphasizes that although technological advances allow a more horizontal search that reveals a larger audience, it is valid to identify the depth of recruitment analysis and it is often necessary to carry out a mixed process, partly online and partly in person, to verify the veracity of the information provided.

For Januzzi (2004), among the disadvantages and limitations of online recruitment are: (1) impersonality: as it is an impersonal contact, the company ends up not having an initial relationship with the candidate and, consequently, does not know him in depth; (2) lack of contact with the company and the candidate: with the internet, the contact is initially virtual, not taking place in person; (3) risk of inappropriate hiring; (4) job offer less than demand: which makes the universe of competing candidates larger.

Companies consider that the technology provided by social networks is a facilitator in the recruitment process, as it allows them to have access to information in an accelerated way, and consequently to have a wide response capacity for people who contact companies (Vieira, 2010).

The author also states that these organizations consider that information is collected through social networks in a more creative and innovative way, providing innovation to companies. In addition to being a very significant factor in stimulating competitiveness, it helps companies to distinguish themselves.

Selection is clearly the least important domain when recruitment and selection companies use the Internet, as it is the least evident objective of Internet use (Vieira, 2010).

7 Final Remarks

This survey is valid for recruiters, candidates and academics who are engaged in recruitment and selection research in the digital age. The advantages and disadvantages of these practices of recruiting and selecting were synthesized and clarified with the help of a bibliographic review, which the results found show that organizations are using the internet as an extremely valuable resource for the execution of recruitment and selection, going according to Furtmuellera et al. (2011), who claim the constant increase in recruiters and candidates in the online environment.

It was identified that the factors, cost and time are the most advantageous in the recruitment and selection procedures in the virtual field, as it becomes favorable for both candidates and company recruiters to access, at any time and update any information without restrictions. In addition to saving time, being a less cost for both.

As disadvantages factors were noticeable the issue of impersonality, the lack of contact and the unreliability of information presented by candidates on platforms, websites and social networks. Suvankulov (2013) states that the positive side is that on platforms and networks, access to information is more detailed. However, Costa (2018) argues that this information lacks veracity.

In view of the above, the research objective was achieved, and the results of the work demonstrated the perspective of several authors in relation to the use of the internet, as well as websites, platforms and virtual social networks in recruitment and selection processes, as well as their awareness. about the possible repercussion in the personal and professional spheres. In addition, a new attitude of care and attention was noted with the information published on these sites.

Finally, it is suggested for future studies the intention of managers to adopt internet platforms, virtual social networks as tools in selection processes, studies to measure the relevance of information found on social networking sites as a source of decision in recruitment and selection processes. As well as a survey on the results of candidates after a recruitment and selection process in the digital environment.

References

Breaugh, J. A., & Starke, M. (2000). Research on employee recruitment: So many studies, so many remaining questions. *Journal of Management, 26*, 405–434.

Caers, R., & Castelyns, V. (2011). LinkedIn and Facebook in Belgium: The influences and biases of social network sites in recruitment and selection procedures. *Social Science Computer Review, 29*(4), 437–448.

Cassiano, C. N., Lima, L. C., & Zuppani, T. S. (2016). A eficiência das redes sociais em processos de recrutamento organizacional. *Navus- Revista de Gestão e Tecnologia, 6*(2), 52–67.

Cembranel, P., Samaneoto, C., & Lopes, F. D. (2013). A inivação das redes sociais virtuais na administração: usos e práticas para a gestão de pessoas. *Revista de Administração e Inovação, 10*(1), 27–50.

Chiavenato, I. (2005). *Gerenciando com as pessoas: transformando o executivo em um excelente gestor de pessoas.* Elsevier-Campus.

Chiavenato I. (1999). *Planejamento, Recrutamento e seleção de Pessoal. Como agregar talentos à empresa* (4.ª Ed.). Editora Atlas SA.

Coradini, J., & Murini, L. (2009). Recrutamento e Seleção de pessoal: como agregar talentos à empresa. *Disciplinarum Scientia, 5*(1), 55–78.

Costa, L. (2012). *Tecnologia da informação na gestão do recrutamento e seleção: a importância do recrutamento on-line.* Monografia [Graduação] Administração Universidade Federal Fluminense. Obtained from: https://app.uff.br/riuff/.../2012-Administração-Livia%20de%20Cássia%20Costa.pdf. Accessed in: April 2019.

Costa, V. (2018). *Utilização da internet nos processos de recrutamento e seleção: uma avaliação de prós e contras à luz do ambiente de negócios de uma empresa do setor de papel e celulose.* Tese de mestrado em Administração pública e de empresas. FGV.

Cunha, M. P., Rego, A., Cunha, R. C., Cabral-Cardoso, C., Marques, C. A., & Gomes, J. F. S. (2010). *Manual de Gestão de Pessoas e do Capital Humano* (2.ª Ed.). Edições Sílabo.

Dhamija, P. (2012). E-recruitment: A roadmap towards E-human resource management. *Journal of Arts, Science & Commerce, 3*(2), 33–39.

Freitas, M. C. O. (2017). *As redes sociais utilizadas como ferramentas do recrutamento das PME de excelência em Lisboa.* Dissertação de Mestrado em Gestão de Recursos Humanos. Universidade Europeia.

Furtmuellera, E. Wilderoma, C., & Tate, M. (2011). Managing recruitment and selection in the digital age: E-HRM and resumes. *Human Systems Management, 30*, 243–259.

Godinho, A. (2009). *E-recruitment - recrutamento e seleção on-line estudo de caso Catho online. Centro Universitário de Brasília Faculdade de Tecnologia e Ciências Aplicadas – Fatecs.* Obtained from: http://www.repositorio.uniceub.br/bitstream/123456789/933/2/20401020.pdf. Accessed in: May 2019.

Guimarães, M. F., & Arieira, J. O. (2005). O processo de recrutamento e seleção como uma ferramenta de gestão. *Revista Ciências Empresariais da UNIPAR, Toledo, 6*(2), 203–214.

Januzzi, L. (2004). *Recrutamento on-line: uma realidade cada vez mais presente nas empresas.* Obtained from: http://www.rh.com.br/ler.php?cod=3739&org=3. Accessed in: March 2019.

Katsurayama, M., Silva, S. R., Eufrázio, W. N., Souza, R. S. A., & Becker, M. A. Á. (2012). Testes informatizados como auxílio na seleção em recursos humanos. *Psicologia: teoria e prática, 14*(2), 141–151.

Kim, S., & O'Connor, J. (2009). Assessing electronic recruitment implementation in state governments: Issues and challenges. *Public Personnel Management, 38*(1), 47–66.

Lima, A. S. H., & Rabelo, A. A. (2018). A importância do e-recrutamento e seleção online no processo organizacional. *Revista Psicologia, Diversidade e Saúde, 7*(1), 147–156.

Lusa. (2015). *Observador.* Obtained from: http://observador.pt/2015/10/02/47-milhoesde-portug ueses-tem-conta-no-facebook/. Accessed in: January 2022.

Marras, J. (2007). *Administração de recursos Humanos.* Futura.

Martins, J. (2007). *Recursos Humanos.* Obtained from: http://w3.ualg.pt/~jmartins/gestao/Final. pdf. Accessed in: April 2019.

Milkovich, G. T., & Boudreau, J. W. (1994). *Human resource management.* McGraw-Hill.

Rajani, S. (2016). A study on recruitment strategies in it progress. *Oncotarget, 7*(21), 31586–31601.

Suvankulov, F. (2013). Internet recruitment and job performance: Case of the US Army. *International Journal of Human Resource Management, 24*(11), 2237–2254.

Vieira, M. P. D. S. (2010). *Impacto das novas tecnologias no recrutamento nas empresas especializadas de recrutamento e seleção.* Dissertação de Mestrado em Gestão de Recursos Humanos. Instituto Universitário de Lisboa.

Zottoli, M. A., & Wanous, J. P. (2000). Recruitment source research: Current status and future directions. *Human Resource Management Review, 10*, 353–383.

Conscious Humanity and Profit in Modern Times: A Conundrum

Ana Martins and **Isabel Martins**

Abstract The aim of the study is to review extant literature so as to evidence that organisations need to develop the individuals, at all levels, in order for innovation and creativity to flourish amidst the current dynamic environment. This development will give rise to innovation and the proposal of new solutions that lead the organisation towards a sustainable and long life. The pioneer theorist, Mary Parker Follett—theorises that, through the concept of constructive conflict, individuals are encouraged to network via the sharing of their experiences and tacit knowledge. Participation at the individual level is fundamental. The Industry 5.0 way of thinking is also highlighted to enhance the human-centred and personalised collaboration evident between humans and machines in this industrial revolution that harnesses the value of innovation to foment sustainability. The study design entails an exploration and critique of extant literature on learning organisations, culture, neoliberalism and entreprenurship within the context of Higher Education Institutions. The value of this study is directed at the Follettian view of integration and entrepreneurship centres in the University space. Research implications and limitations entail that the study could be further developed by conducting primary data collection in order to ascertain the transition from the neoliberal to the Follettian model.

Keywords Creativity · Culture · Entrepreneurship · Learning organisation · Leadership · Innovation · Sustainability · University

A. Martins (✉)
Graduate School of Business and Leadership, University of KwaZulu-Natal, Westville, South Africa
e-mail: MartinsA@ukzn.ac.za

I. Martins
School of Management, IT and Governance, University of KwaZulu-Natal, Westville, South Africa

1 Introduction

This study sets forth a review of extant scholarship with the view to evidence how organisations require to innovate and create in order to be sustainable and long-living. The chapter is divided into the following areas: a contextualization of the fifth industrial revolution; followed by the theoretical framing of constructive conflict as Mary Parker Follett posited. The constructs of learning and creative co-leadership as well as learning and innovation arising through critical awareness. A critique of the neoliberal paradigm is provided as this is directed at the notion of entrepreneurship as being the panacea albeit this has led to the perpetuation of the problems associated with capitalism. This study also provides its limitations and indicates directions for future research avenues.

2 Human Centred Industrial Paradigm

Follett theorises that universities should advocate education to concentrate on learning which is intergenerational in nature and directed at achieving the good for society as opposed to personal good (Follett, 1970). The goal of the university is grounded on the principles of a democratic society which entail to educate learners with critical thinking skills and also the development of character which are considered to be as the cornerstone for a society based on democracy, according to Follett (1918). Indeed, the educator and the student should develop a relationality committed to interrelate theory with practice in order to fully promote the improvement of society (Follett, 1918). This will enable organizational learning to occur as a result of assimilating absorptive capacities—learning from the external environment. Follett further purports the notion associated with the "law of the situation" (Follett, 1919, 1941), or a distinctiveness that other individuals are able to take part in. Therefore, the learner and teacher become bonded in the pursuit of encouraging learners to use their imagination (Follett, 1970; Wheelock and Callahan, 2006) in order to achieve a mutual/shared goal.

After the second world war, education went from being only for the elite (Scruton, 2015) to being available for the masses (Noble, 2012). Indeed, knowledge became a good for public consumption (Stiglitz, 1999) thus evidencing a culture of learning based on inclusivity. In the 1970s, public funding for universities started to decline and this brought about the rise in student debt and education was no longer considered a public commodity at the core of democracy. This defunding tendency to control the university is a form of neoliberalism. As a result, this has given rise to the concept of academic capitalism wherein education products and related-objects and the intellectual property (academics as well as all learning materials) are attributed a market value. Additionally, economic crises have compounded this context and the initial democratic model with its collegiality, has been replaced by the hierarchical structure as prescribed by governance and the associated control mechanisms, namely, teaching workloads and plans, performance appraisals and monitoring, as well as research outputs.

Industry 5.0 has arisen from a sequence of previous industrial revolutions, namely the first industrial revolution also known as Industry 1.0, the eighteenth century, heralded the steam power invention thus bringing the mechanisation of power. In the nineteenth century, electricity was invented in the second industrial revolution and this included the assembly line production. Industry 3.0 refer to the third industrial revolution was initiated in the 1970s, and introduced partial automation, computers and mass production. Currently, the 4th industrial revolution, also termed as Industry 4.0, denotes the digitalization pertaining to manufacturing; smart cities and networking. In the twenty-first century, the 5th industrial revolution, also known as Industry 5.0, and is anticipated to bring humans and machines together through personalization of the experience by synchronising technology with human thinking capabilities thus enhancing collaboration. This revolution seeks to achieve and enhance the added value of sustainability.

Carayannis et al. (2020, p. 2) posit that "Industry 5.0 is considered to be the answer to the question of a renewed human centered/human centric industrial paradigm, starting from the structural, organizational, managerial, knowledge-based, philosophical and cultural reorganization of the production processes of industry". Nahavandi (2019, p. 3) indicated that "the Fifth Industrial Revolution will pair human and machine to further utilize human brainpower and creativity to increase process efficiency by combining workflows with intelligent systems. While the main concern in Industry 4.0 is about automation, Industry 5.0 will be a synergy between humans and autonomous machines". The current knowledge economy provides the possibility to manage the opportunity that arises from complexity and ambiguity by channelling innovation, knowledge sharing and creation in smart spaces. This type of candour is aligned to the innovative Industry 5.0 way of thinking wherein answers for sustainable growth emphasise humans as being at the core.

3 Learning and Creative Co-leadership

The notion of co-leadership entails capabilities and characteristics which are very similar to distributed leadership, viz-à-viz, cultivating dialogue, versatility, candidness, and promoting a culture based on innovation (Martins and Martins, 2022). The basic principle inherent in viewpoints that may appear to be divergent yet converge, present an opportunity as the situation may give rise to the design of creative ideas through the harmonising effect stemming from the integration of the different ideas. Follett theorised that conflict is related to difference and should be channelled productively arising from teams which are extremely diverse in nature. What is considered fundamental is the interaction the individual has with the context within which the individual is immersed. If domination and compromise prevail, then innovation is not achieved as only temporary gains arise instead of achieving a collective good. Additionally, inventive integration arises when individuals listen attentively to the various different ideas. Furthermore, collective good ensues when the organisation endeavours to achieve fundamental conditions which include an open-minded attitude as well as cooperative thinking. Indeed, integrating conflict presupposes "a high degree of intelligence, keen perception and discrimination, (and) more than all, a brilliant inventiveness" (Follett, 1925 in Metcalf, 1941). What is more, it is presupposed that individuals are realistic, logical, sensible, reliable, are not self-conceited and get on well with others. This harnesses an environment conducive for networking which is a form that contemporary organisations have embraced in the current knowledge economy and society. Pfeffer (1990) theorises that these network-type organisations highlighting the characteristics of teams which include empowerment of members, participation and sharing of information.

The context within which individuals can interact arises from the space wherein sharing of their experiences related to tacit knowledge and which are experiential and intangible in nature, as well as explicit knowledge which is tangible. Managers and leaders should foster opportunities for individuals to communicate and build a rapport with one another in order to overcome indifference and instead to promote a sense of the common goal and intention. Follett further posits that 'power-with' is developed over time and is closely related to empowerment of the individual which is more favourable in comparison to power over which refers to the control of others (Carlsen et al., 2020). Additionally, power-with is aligned to experience and knowledge. Indeed, leadership is context-driven and Follett further corroborates that the dynamic and circular interaction between individuals is the trust-building context in which innovative ideas can arise in order to solve complex situations. Moreover, the Folletian view of collaboration and power-with as opposed to the Fayolian view of bureaucracy and power-over, the human element prevails and this makes it possible for individuals to be innovative and creative (Martins and Martins, in press). Learning is considered essential and in this regard, Follett (1924) further postulates three core constructs embedded in five principles to endorse a circular adaptive learning process. The three core concepts entail: (1) reasoning in the form of the entire context and not odd parts thereof; (2) channelling the imaginative and resourceful capacity embedded in the notion of integrative thinking which is shaped by the individual; and (3) boosting one's capability to act in accordance with the situation. The five principles entail: (1) engaging one's inner awareness of intent and aim, purpose and principle; (2) engaging with others who are different and acquiring knowledge arising from the purposes, beliefs, ideals, values; (3) constructing change by engaging in a co-creation process resulting from the differences in (2) above; (4) endorsing and working on blueprints which support experiences; and (5) familiarize oneself with and learning from the aforementioned experiences, which includes the actions of doing, learning about the particular subject, as well as actually living life. Follett considers the awareness of consciousness as well as accountability to be fundamentally underpinning the adaptive learning circle.

The notion of constructive conflict is further substantiated by Tjosvold (1988), Leonard and Straus (1997), as well as Jameson (1999). For Pascale (1990), creativity arises from discord and contradiction, commitment, eagerness. Swanson and Holton (2001, pp. 145–146) posit that learning as well as development are viewed "as avenues to individual growth; a belief that organizations can be improved through learning and development activities; a commitment to people and human potential; a deep desire to see people grow as individuals; a passion for learning". This view is further corroborated by Gilley et al. (2002). Follett's notion of a leader is characterised as an individual who shares this authority with other individuals in the organization (Follett, 1987, 1998/1918). "Authority follows the function... belongs to the job and stays with the job" (Follett, 1996, p. 153). Indeed, Follett (1998/1918) theorises learning as being a perpetual activity throughout life. Lindeman (1989/1926) considers that "education is life" (p. 129) which is in harmony with the sentiments of Follett.

With the changes that have come about in the current global economy and society, the learning organisation characteristics have also moved on to be viewed as the following: personal mastery is considered to be aligned with the mindset of lifelong learning and is directly in tune with the learning culture. This mindset should prevail across the entire organisation and its ecosystems which nurture the systems thinking approach through the collaborative learning culture. The characteristic of mental models can be viewed as the propensity to innovate the organisation corroborated by a leadership that is focused on the future in its shared vision. All the above is also strengthened through knowledge sharing and learning, at the team level. As much as it is important to support the notion of the learning organisation, it is also vital to be in some way against it. The will to learn also comes with the resistance to learn. The latter entails cultures that are toxic, organisations that avoid reflection.

This is further harmonised with Fayol's theory of management. The concept of creative leadership is theorised in the notion that the "Fayolian leader is able to launch ventures, generate motivation, and develop the collective capability to face an unpredictable future productively. Fayol theorised a new type of authority that we call 'creative'" (Hatchuel & Segrestin, 2019, p. 407). Indeed, Fayol postulates the notion of "management of innovation" (Hatchuel & Segrestin, 2019, p. 408) with the central tenet of management being *"perfectionnement"* (Hatchuel & Segrestin, 2019, p. 409). Planning is considered as foreseeing and being prepared for the unpredictable circumstances which also encompasses good organising by taking into account the unknown. The contingency approach that Fayol theorises is even more pertinent in the current globalised world of rampant change and flux. Being united and flexible, enables the individuals in the organisations to handle the unknown as opposed to the uncertain. Managers, who are vested with trust by all employees, are expected to provide indication as opposed to plans.

Fayol "outlines the creative aspects of the design of programmes. Building the programme can use existing models but must remain open to creative invention. The unknown requires both freedom of action and creative capability" (Hatchuel & Segrestin, 2019, p. 406). Lambert (2022) reviewed the work of March and highlighted that being restricted by a long-term strategy is not always conducive to fostering individual learning (as posited by Cyert & March, 1963/1992); instead what is appropriate for current organisations is to emphasise short-term reactions as well as experiences and their related feedback. Indeed, individuals become open and receptive to learning as a result of their experiences. Additionally, trust is a fundamental component in the learning context especially in situations where ambiguity is rife as trust enables the individual to feel integrated and part of the organisation. Alienation occurs when mistrust is prevalent between individuals in the organisation.

4 Critique of the Neoliberal Paradigm

It has become apparent that entrepreneurial education encourages neoliberalism in this sector as this can lead to societal inequalities. Neoliberalism enables the self-organization of individuals, groups and organisations in society to achieve well-being. Neoliberalism entails seven features, namely, (i) state deregulation, (ii) privatisation, (iii) regulation that is customer oriented, (iv) marketisation, the (v) utilisation of market representatives in government sectors, (vi) Non-Governmental Organisations are supported to undertake citizenship as well as (vii) to encourage individuals to become self-supporting and independent. Therefore, on the positive side, neoliberalism amalgamates the opportunity for the individual to have the freedom of choice to channel their expertise within the most favourable utilisation in order to achieve superior echelons regarding health, wealth, general well-being as well as efficiency in society as opposed to that which other types of economic policies could propose (Lackéus 2017, Robinson, 2010; Rose, 1999). The negative side is that neoliberalism steers towards the maximisation of profits, common good and the needs of the people are disregarded, values are placed on profit-making and that which is against democracy. Hence, it has become apparent to handle entrepreneurship with care. Entrepreneurial education is being considered as a way to reduce various demands arising in society, as well as at the level of the individual. The latter regards entrepreneurial education as affording the opportunity for the individual to be independent, resourceful, displaying an attitude that is ingenious, innovative and creative both to work and life itself. This enables the individual to be constantly searching for opportunities. On the societal level, for example, to improve the growth of the economy; advance those competencies deemed as crucial; to improve the level of student involvement which leads to completion and throughput. What has been evident is that values based on capitalism have arisen as a result of entrepreneurial education.

Entrepreneurship entails a positive focus, as Morris et al. (2012, p. 208) substantiate, which resides in entrepreneurship education being regarded as a route to achieving a life with meaningfulness as opposed to the conventional route for generating wealth. The notion of "students-as-givers engenders the responsible citizenship on the individual as opposed to student-as-taker" (Lackéus, 2017, p. 646). A culture centred on the human beings fosters improved organisational performance and higher levels of motivation. Additionally, the sharing and creation of knowledge are greatly diffused in the organisation (Cillo et al., 2022).

Neoliberalism was implemented in the 1970s and is associated with the range of politico-economic ideas prevailing at the time; this is also intimately linked to the cultural change that was initiated at that time. The so-called 'free market' was at the base of this neoliberalism. The latter encompasses the liberalisation of trade, privatisation of public services, the state became deregulated and decreased in size which is designated to be "politically assisted market rule" (Peck, 2010, p. xii). Neoliberalism with its irresponsible deregulation of rules and standards set out by the market, is closely linked with dishonest procedures followed by entrepreneurs.

The underlying tenet of neoliberalism is its power to control the consciousness/mind of the individual without apparently penalising the physical body; in other words, the physical body has been allowed all freedom in this so-called democracy but what is being controlled, is the mind. This neoliberalism poses as a benevolent system in that it does not impose anyone to do anything directly. Instead, it invokes values pertaining to free expression but which in reality are a means of control *"in the name of freedom"*—which is "the autonomous individual 'free to choose'" (Rose, 2017, p. 304). This neoliberalism therefore, has instilled in the individual the notion of self-reliance, whereby the state has relinquished all responsibilities and these are now solely in the hands of the individuals. This self-reliance of individualism was taken to the extent that the state is no longer responsible for job creation; instead this lies in the hands of the individuals. This self-reliance has created a culture of "hatred for dependency" (Solnit, 2018, p. 46). Even though neoliberal freedom is intimately associated with the market, and the vast surplus of products offered by the market, this is considered a paradox. According to Marttila (2012, p. 5), "the neoliberal role model of social subjectivity" is a synonym for the entrepreneur.

Currently, entrepreneurship and indeed its mandatory inclusion within the HEIs curricula, across all universities globally, has gained importance. Moreover, entrepreneurship is in line with the current dominant economic discourse, i.e., that of neoliberalism. According to Dardot and Laval (2013, p. 103) reasoning embedded in neoliberalism endeavours "to shape subjects to make them entrepreneurs capable of seizing opportunities for profit and ready to engage in the constant process of competition". The authors of this chapter are, therefore, of the opinion that neoliberal discourse on entrepreneurship with its perpetuation of capitalistic, market orientation, and its destructive underlying principles, is the essence of the problem which it claims to address. Entrepreneurship was engineered to rescue capitalism thus allowing capitalism to mercilessly continue with its direction of increasing production and also surging consumption. The current discourse that dominates entrepreneurship is based on neoliberalism as these debates feed this economy, wherein the public sector has been subjected to downsizing and deregulation on the one hand, and on the other hand, the accountability of the individual has gained ground, Jessop (2017). According to Rosile et al. (2013), a new model for public universities can be found in the principles which Follet postulated as the ensemble learning theory (ELT) and entrepreneurship centres. This ELT model, is situated in the presence of constructive conflict (Follett, 1919, 1941). The authors of this chapter are of the opinion that this model could be applied to contemporary universities instead of the neoliberal economic model which is currently being applied. Follett's model envisages entrepreneurship centres, the space where integration occurs between students, academics, administrators and the public. As a result of integration, all these stakeholders are in alignment with a reciprocal, communicated and collective activity (Follett, 1919). Furthermore, this Follettian view (Novicevic, 2007) entails two distinct fields namely, education and economy, which cease to exist in isolation and become integrated to ensure prosperity is channeled to the community. As Nelson (2017) corroborates, this context engenders innovation and creativity.

5   Conclusion

In this book chapter, the Follettian notion of integration has been highlighted to evidence the constructs of power-with and constructive conflict in order to underpin the notion of creative co-leadership as being the conduit to innovation and creativity. The narrative then emphasized a critique of the neoliberal paradigm in entrepreneurship, as prevailing in contemporary universities. Thereafter, the narrative developed with the Follettian notion of entrepreneurship centres and ELT that should be applied as opposed to the current neoliberal approach.

References

Carayannis, E. G., Dezi, L., & Gregori, G. (2022). Smart environments and techno-centric and human-centric innovations for industry and society 5.0: A quintuple helix innovation system view towards smart, sustainable, and inclusive solutions. *Journal of the Knowledge Economy, 13*, 926–955.

Carayannis, E. G., Draper, J., & Bhaneja, B. (2020). Towards fusion energy in the industry 5.0 and society 5.0 context: Call for a global commission for urgent action on fusion energy. *Journal of the Knowledge Economy, 12*(4), 1–14.

Carlsen, A., Clegg, S., Pitsis, T., & Mortensen, T. (2020). From ideas of power to the powering of ideas in organizations: Reflections from Follett and Foucault. *European Management Journal, 38*(6), 829–835.

Cillo, V., Gregori, G. L., Daniele, L. M., Caputo, F., & Bitbol-Saba, N. (2022). Rethinking companies' culture through knowledge management lens during Industry 5.0 transition. *Journal of Knowledge Management, 26*(10), 2485–2498.

Cyert, R. M., & March, J. G. (1963/1992). *A behavioral theory of the firm* (2nd ed.). Prentice-Hall, Blackwell Business.

Dardot, P., & Laval, C. (2013). *The new way of the world: On neoliberal society* (G. Elliott, Trans.). Verso.

Follett, M. P. (1919). Community is a process. *The Philosophical Review, 28*(6), 576–588.

Follett, M. P. (1924). *Creative experience*. Longmans: Green.

Follett, M. P. (1925). Constructive Conflict. In P. Graham (1995) Mary Parker Follett—Prophet of Management (pp. 67–87), Boston, MA: Harvard Business School Press, Reprinted from E. M. Fox & L. Urwick (Eds.), 1973, Dynamic Administration: The Collected Papers of Mary Parker

Follett (p.120), London: Pitman. This paper was first presented before a Bureau of Personnel Administration Conference Group in January 1925.

Follett, M. P. (1941). *Dynamic Administration: The Collected Papers of Mary Parker Follett*, edited by Metcalf, H. C., & Urwick, L. F. London, NY: Harper and Brothers.

Follett, M. P. (1970). The teacher-student relation. *Administrative Science Quarterly, 15*(2), 137–148.

Follett, M. P. (1987). The illusion of final authority. In L. Urwick (Ed.), *Freedom & coordination* (pp. 1–15). Garland.

Follett, M. P. (1996). The basis of authority. In P. Graham (Ed.), *Mary Parker Follett—prophet of management* (pp. 141–162). Harvard Business School Press.

Follett, M. P. (1998/1918). *The new state-group organization. The solution for popular government.* The Pennsylvania State University Press.

Gehani, R., & Gehani, R. (2007). Mary Parker Follett's constructive conflict: A psychological foundation of business administration, for innovative global enterprises. *International Journal of Public Administration, 30*(4), 387–404.

Gilley, J. W., Eggland, S. A., & Maycunich-Gilley, A. (2002). *Principles of human resource development* (2nd ed.). Cambridge, MA: Perseus.

Hatchuel, A., & Segrestin, B. (2019). A century old and still visionary: Fayol's innovative theory of management. *European Management Review, 16*(2), 399–412.

Jameson, J. K. (1999). Toward a comprehensive model for the assessment and management of intra-organizational conflict: Developing the framework. *The International Journal of Conflict Management, 10*(3), 268–294.

Jessop, B. (2017). Varieties of academic capitalism and entrepreneurial universities. *Higher Education, 73*(6), 853–870.

Lackéus, M. (2017). Does entrepreneurial education trigger more or less neoliberalism in education? *Education + Training, 59*(6), 635–650.

Lambert, G. (2022). James March: A postmodern perspective on organization without management theory. *Journal of Management History, 28*(1), 66–88.

Leonard, D., & Strauss, S. (1997). Putting your company's whole brain to work. *Harvard Business Review*, 111–121.

Lindeman, E. C. (1989/1926). *The meaning of adult education*. Norman, OK: Harvest House.

Martins, A., & Martins I. (2022). In favor of leaderless management: Follettian perspective of co-leadership. In F. Hertel, K. M. Jørgensen, & A. Örtenblad (Eds.), *Debating leaderless management. Can employees do without managers? Palgrave debates in business and management book series* (1st ed.). Palgrave Macmillan Publishing.

Martins, A., & Martins, I. (in press). Seeking new ground: Quark influence on the Transhuman future. In For the forthcoming book, C. Machado & J. Paulo Davim (Eds.), *Managerial challenges of industry 4.0*, Edition Diffusion Presse Sciences—EDP Sciences.

Marttila, T. (2012). *The culture of enterprise in neoliberalism: Specters of entrepreneurship.* London: Routledge.

Metcalf, H. C., & Urwick, L. (Eds.). (1941). *Dynamic administration: The collected papers of Mary Parker Follett*. Harper and Brothers.

Morris, M. H., Kuratko, D., Schindehutte, M., & Spivack, A. (2012). Framing the entrepreneurial experience. *Enterpreneurship Theory Practice, 36*(1), 11–40.

Nahavandi, S. (2019). Industry 5.0–a human centric solution. *Sustainability, 11*(16), 43–71.

Nelson, G. (2017). Mary Parker Follett—Creativity and democracy, human service organizations: Management. *Leadership & Governance, 41*(2), 178–185.

Noble, D. F. (2012). *Digital diploma mills: The automation of higher education*. Aakar Books.

Novicevic, M., Harvey, M., Buckley, M., Wren, D., & Pena, L. (2007). Communities of creative practice: Follett's seminal conceptualization. *International Journal of Public Administration, 30*(4), 367–385.

Pascale, R. T. (1990). *Managing on the edge: How the smartest companies use conflict to stay ahead*. New York: Simon & Schuster.

Peck, J. (2010). *Constructions of neoliberal reason*. Oxford University Press.

Pfeffer, J. (1990). Producing sustainable competitive advantage through the effective management of people. *Academy of Management Executive, 9*(1), 55–72.

Robinson. (2010). *Guardians of the System? An Anthropological Analysis of Early 21st Century Reform. of an Australian Educational Bureaucracy.* Doctoral Thesis, University of Western Australia.

Rose, N. (1999). *Governing the soul: The shaping of the private self*. Free Association Books.

Rose, N. (2017). Still "like birds on a wire"? Freedom after neoliberalism. *Economy and Society, 46*(3/4), 303–323.

Rosile, G. A., Boje, D. M., Carlon, D. M., Downs, A., & Saylors, R. (2013). Storytelling diamond: An antenarrative integration of the six facets of storytelling in organization research design. *Organizational Research Methods, 16*(4), 557–580.

Scruton, R. (2015). The end of the university. *First Things, 252*(6), 25–30.

Solnit, R. (2018). *Call them by their true names: American crises (and essays)*. Granta Books.

Stiglitz, J. E. (1999). Knowledge as a global public good. In I. Kaul, I. Grunberg & M. Stern (Eds.), *Global Public Goods: International Cooperation in the 21st Century,* (Vol. 308, pp. 308–325) Oxford: Oxford University Press.

Swanson, R. A., & Holton, E. F. (2001). *Foundations of human resource development*. Berrett-Koehler.

Tjosvold, D. (1988). Putting conflict to work. *Training and Development Journal*, 61–64.

Wheelock, L., & Callahan, J. (2006). *Human Resource Development Review, 5*(2), 258–273.

Multigenerational Men and Women and Organisational Trust in Industrial Multinational Firms in Portugal

Lurdes Pedro and José Rebelo

Abstract The concept of organisational trust, has been considered as a promoter of increased performance and, as such, has raised a broad and growing interest in the organisational literature. However, there are other less studied variables with apparently positive links, which have not been consistently confirmed by empirical research and therefore a more thorough understanding is required. The study was carried out in four multinational industrial companies with the primary objective of analysing the relationship between organisational trust, service length (different generations in organisations) and employee gender. For this purpose, a questionnaire survey was used — Schoorman and Ballinger's scale (Leadership, trust and client service in veterinary hospitals. Purdue University, 2006)—which was applied in these four organisations and continues to be one of the most promising instruments for the study of trust between subordinates and leaders or managers. The study provides a set of results that characterize the degree of organisational trust, showing not only that trust is slightly lower among women when compared to men in these organisations, but also that is higher among employees with less service length in the company. In conclusion, the article outlines implications for practice and fosters further discussion and future research.

Keywords Multigeneracional Employees · Human resources management · Organisational trust

1 Introduction

Organisational Trust is one of the management topics and it is a widely studied specific area in academic researches. Organisational Trust has raised a growing

Present Address:
L. Pedro · J. Rebelo (✉)
Escola Superior de Ciências Empresariais, Instituto Politécnico de Setúbal, Setúbal, Portugal
e-mail: Jose.rebelo@esce.ips.pt

L. Pedro
e-mail: lurdes.pedro@esce.ips.pt

interest due to its great importance in reinforcing organisational culture and, above all, because it is a common assumption that it plays a fundamental role in employee performance levels.

This research, about trust, was conducted on employees of several generations in the largest multinational industries in Portugal. The main purpose of the study was to examine the relationship between organisational trust, service length (different generations in organisations) and employee gender and its characterisation.

For this purpose, we opted for a quantitative methodology through the application of a questionnaire to the employees of four multinational companies in the electrical and electronic sector. The information obtained was then subjected to association and correlation analyses.

The structure of this study is composed of three sections: (1) literature review; (2) methodology; (3) analysis and discussion of results.

2   Literature Review

2.1   Theoretical Framework

Trust has recently became a central topic of study by researchers (e.g., Kramer, 2012; Rotter, 1967; Schoorman et al., 2007) in order to understand the mechanisms through which people trust and how they shape social relationships accordingly.

The increasing interest in the study of trust, in these last decades, has led to the emergence of different perspectives, particularly in the scientific areas of sociology (Granovetter, 1985), economics (Williamson, 1993), anthropology (Uslaner, 2002), psychology (Webb & Worchel, 1986) and human resource management (Veloso & Pinto, 2021).

This increasing interest in the study of trust arose from the premise that trust can play a core role not only in organisational dynamics and in the success of organisations, but also in a fundamental element of organisational and social performance, in the stability of social relations and economic prosperity (Fukuyama, 1995; Lewis & Weigert, 1985).

Trust has become an urgent and core concern in the current context in particular because of the role it plays in organisational productivity and individual performance (Colquitt et al., 2007; Lewicki et al., 1998; Mayer & Davis, 1999; Mayer et al., 1995). It emerges as a necessary resource, working as a mechanism upon which actors establish simpler interactions, with a reduced level of monitoring and control demands among people in the organisation and among organisations, due to a belief in the credibility of a person or system.

This coordination mechanism among the actors, makes operations more efficient and faster (Kramer, 2012).

Trust stimulates cooperation and effective communication, contributing to the organisations' success, not only in a short-term but also in a long-term perspective

(Mishra, 1996; Whitener et al., 1998). It is an important factor for the organisation stability and employee well-being (Hendriks et al., 2020). It is a way to ensure cooperation between people with different interests (Hasche et al., 2022).

According to this perspective, trust refers to a set of beliefs, a psychological state of assurance that the employee feels towards his/her company, especially in situations of uncertainty and confrontation that may risk in his/her relationship with his/her organisation.

Trust is a multidisciplinary concept and has been studied at multiple analysis levels. Although the study of trust in different scientific areas has strengthened and broadened its interest in literature, it has also brought about a multiplicity of meanings in the concept's definition (Hosmer, 1995). It is considered as a construct, with different perspectives converging in its understanding as a multidimensional phenomenon (Rodrigues & Veloso, 2013).

One the one hand trust can exist at an organisational level (that of a employee in his/her organisation) (Colquitt et al., 2007; Zaheer et al., 1998). One the other hand there is also an interpersonal level (Hassan & Semercioz, 2010), when the focus is on the individual, that is, when it develops among employees/co-workers or between employees and supervisors.

Organisational trust is the generalisation of the mutual trust model among individuals, which includes not only employee's trust in his/her organisation, but also each member's trust in his fellow worker, this resulting in a more effective collaboration between the elements of the organisation (Katou, 2013).

Organisational trust is connected to the maintenance of the psychological contract (Robinson, 1996), by being perceived as essential in the individual's interaction in the organisation and in its stability.

The respect or violation of the psychological contract by the organisational actors is therefore likely to increase or decrease the level of employee trust in his/her organisation and vice versa (Robinson, 1996; Rousseau et al., 1998).

There have been several definitions of organisational trust, which are summarised in Table 1.

With the increasing number of publications, in the various fields of knowledge, different approaches in the mid-1990s (*i.e.*, Mayer et al., 1995; Rousseau et al., 1998) tried to understand the phenomenon of building trust in organisations and to identify the elements and relationships involved in all the intervenients.

The trust model developed by Mayer et al. (1995) is one of the most widely used models in research. It emerges as the willingness of an individual to place himself in a vulnerable position in a relationship with another person or group. According to these authors, trust involves the predisposition to vulnerability and implies risk acceptance in trust-based relationships, contributing to increase the organisational effectiveness through greater reciprocity and the presence of less complex relationships.

Rousseau et al. (1998: 395) added another idea to this construct of trust which is *"a psychological state that includes the willingness to place ourselves in a situation of vulnerability vis-à-vis another person, based on positive expectations about their*

Table 1 Definitions of trust in literature

Author/year	Definitions
(Rotter, 1967)	Trust is an expectancy held by an individual or group that can rely upon the word, promise, verbal or written statement of another person or group
(Kee & Knox, 1970)	In the simplest and perhaps the most common case, a trust situation involves two parties which are to a certain extent interdependent with respect to the outcomes defined by their joint choices, and one of the parties (P) is confronted with the choice between trusting or not trusting. However, P's choice not to manifest trust toward O will preclude betrayal, leaving O usually with no further option with respect to the particular situation. It is noteworthy that both P and O are aware of the risk to which P exposes himself in his decision to trust O. ("Risk" here refers to the possibility that O can-but not that he necessarily will-betray P's trust.) That is, P knows O can betray him and O knows that P has extended his (P's) trust even in the face of that risk. Therefore, even where the risk is perceived as negligible, the situation still involves trust, as long as O's betrayal is a possibility
(Barber, 1983)	The first of these two specific definitions is the meaning of trust as the expectation of technically competent role performance. [...] The second meaning of trust that I shall analyze concerns expectations of fiduciary obligation and responsibility, that is, the expectation that some individuals in our social relationships have moral obligations and should therefore demonstrate a special concern for others' interests above their own.
(Lewis & Weigert, 1985)	In terms of behaviour, to trust is to act as if the uncertain future actions of others were indeed certain in circumstances wherein the violation of these expectations results in negative consequences for those involved. In other words, the behavioral content of trust is the undertaking of a risky course of action on the confident expectation that all persons involved in the action will act competently and dutifully
(Mayer et al., 1995)	[...] the willingness of a party to be vulnerable to another party's based on the expectation that this party will take a particular course of action, which is important to the trustor, regardless of this ability to monitor or control that other party
(Rousseau et al., 1998)	Trust is a psychological state comprising the intention to accept vulnerability based upon positive expectations of the intentions or behavior of another
(Hardin, 2002)	Trust is seen as an intentional relation, the rational analysis of which must depend on the rationality of both the trustor and the trustee and on the commitments of the trustee
(Bhattacherjee, 2002)	Trust is the expectation of positive or non-negative outcomes which derive from an expected action from another party and it is characterised by uncertaintly, meaning trust related to good results

(continued)

Table 1 (continued)

Author/year	Definitions
(Saparito et al., 2004)	Relational trust refers to a "trustor's" confident belief that a "trustee" will act beneficially because the trustee cares about the trustor's welfare
(Krishnan et al., 2006)	Building on this prior research, we define interorganizational trust as the expectation held by one firm that another will not exploit its vulnerabilities when faced with the opportunity to do so […]
(Six et al., 2010)	[…] we define trust as a psychological state comprising the intention to accept vulnerability to the actions of another party based upon the expectation that the other will perform a particular action that is important to you person
(Bozic et al., 2019)	Trust is a psychological state comprising the intention to accept vulnerability based upon positive expectations of the intentions or behaviour of another person

Source Adapted from Oliveira et al. (2020) and Pedro (2015)

intentions and behaviour". Although these definitions are different, there are core elements that helped to support the development of trust in a relationship: risk and interdependence among the parties (e.g., Schoorman et al., 2007).

Trust implies placing hope and positivity in relationships, which reduces the discomfort of uncertainty and risk perception. It is an adaptive process linked to human nature (Bering, 2010) and the search for balance.

There is an acceptance of risk and uncertainty when people interact with one another and believe that these interactions produce positive outcomes.

When two people feel they can believe and trust each other, a sense of trustworthiness develops between them. Trustworthiness corresponds to the attributes perceived by others, and have proven to be an anchor for the one who trusts and is available to accept vulnerability (Barczak et al., 2010). Thus, the willingness to trust is, by itself, insufficient for the establishment of a trust relationship.

It is also based on the characteristics of those who trust, which are used by some authors as a personality variable (e.g., Mayer et al., 1995). From their experiences of trust, positive or negative, people tend to extrapolate these experiences and construct beliefs or develop general expectations about others. Rotter (1967) reinforces this perspective by arguing that individuals tend to acquire a diffuse expectation to trust others according to the individual personality characteristics.

The propensity to trust does not depend exclusively on the various experiences of interaction between two or more people, it is dependent on dispositional factors to trust, which are linked to personality, therefore being a relatively stable individual capacity (e.g., Rotter, 1967).

Schoorman et al. (2007) revision of the trust model integrates new analysis strands such as unidirectionality because trust is not necessarily mutual, nor reciprocal, but also emotional and as such it affects the impact analysis on trust or the impact of trust violation.

They also proposed the need to specify the contextual variables that lead to the understanding of trust propensity and the salience of trust variation across different cultures. Notwithstanding the model revision, the core elements of Mayer et al. (1995) proposal are maintained, as well as vulnerability and the belief in positive expectations towards the other which also prevail.

2.2 Antecedents, Consequences, Mediators and Moderators of Trust

Several empirical studies have suggested that trust increases cooperation and team-work, improves communication and employee satisfaction, creates more positive attitudes, facilitates organisational citizenship behavior, and increases the perfor-mance of individuals, groups, and organisational performance (e.g., Davis et al., 2000; De Jong et al., 2016; Dirks & Ferrin, 2001; Matzler & Renzl, 2006; Podsakoff et al., 1990).

In leader's trust meta-analysis developed by Dirks and Ferrin (2002), it is evident that subordinates' trust is an essential component of effective leadership. Mayer and Gavin (2005) concluded that trust in senior managers makes employees focus more on tasks that add value to the organisation and that trusting these leaders is more related to organisational citizenship behaviors than to individual performance.

Trust has emerged as a mediator of a relationships set, as a relationships facili-tator between various management elements, such as information sharing, motiva-tion, satisfaction, conflict reduction and work environment outcomes, influencing a person's expectations about other's future behavior (Dirks & Ferrin, 2002).

Trust has also emerged as a moderator, in the interaction between those who trust and those who are trusted, influencing the responses to action and the perceptions of those who trust. These studies have related the positive effects of trust, namely in the relationships between leaders and followers (Dirks, 2000; Dirks & Ferrin, 2002; Dirks & Skarlicki, 2009).

2.3 Individual Characteristics in the Relationship with Organisational Trust

Some studies (e.g., Zak & Knack, 2001) try to explain trust with the feeling of belonging to groups, as a cultural system based on the expectations shared by the groups. Empirical evidence has not corroborated this perspective as it didn't confirm lower trust levels in more heterogeneous and unequal societies, or higher in more homogeneous societies (Guinot & Chiva, 2019).

Garbarino and Slonim (2009), Zucker (1986), Mahdizadeh and Hosseini (2010) (cit in Guinot & Chiva, 2019) and Maddux and Brewer (2005), confirmed that the trust varies according to gender, age, service length in the organisation and other individual characteristics. Our work is an extention of this literature as it documents how gender and service length in the organisation, may be a factor in interpersonal trust.

3   Methodology

The study aims at verifying the relationship between organisational trust, as defined in the reference literature and the variables gender and service length in the organisation. The current research is a quantitative study. For the operationalisation of this study, we used the questionnaire developed by Mayer and Davis (1999) and Mayer and Gavin (2005) which was revised in a shorter 7-item scale by Schoorman and Ballinger (2006). The items were rated on an adjusted 6-point Likert scale (*1-Strongly Disagree and 6-Strongly Agree*), instead of the authors' proposed 7-point scale.

This questionnaire was applied in four multinational companies in the industrial electrical and electronics sector, which was considered as a convenience sample. A total of 511 valid responses were obtained from the four organisations.

Descriptive statistics were used, as the use of a convenience sample does not allow generalisation of the study results (Pestana & Gageiro, 2014).

3.1   Data Collection Procedure

Schoorman and Ballinger's (2006) trust scale was validated for the Portuguese reality with good results in terms of reliability and validity. This scale remains one of the most promising trust scales for its psychometric properties and was built from the original scale with 7 items, while trying to preserve the conceptual definitions but mitigating redundancies of their meanings (Schoorman et al., 2007).

This scale measures the extent to which employees trust their superiors to make decisions and are open to criticism from them, as well as to taking risks regardless of their ability to monitor or control them. This is based on prior knowledge about the perceived integrity, benevolence and competence of superiors and on positive expectations that their actions are always well-intentioned.

The questionnaire includes demographic individuals variables created by the researchers. There are certain variables such as gender, age, educational qualifications, and service length in the company, among others.

The study was conducted in four organisations from the industrial electrical and electronics sector.

They are multinational companies operating in Portugal, located in the industrial cluster of Lisbon and Braga, of Swiss, North American and French origin and are on the list (INE, 2022) as the 50 largest companies in the sector in terms of turnover and number of employees.

SPSS 28 (statistical package for the social sciences software) statistical software is employed in the conducted analyses. Spearman's Ró, ANOVA, and bivariate analysis, association and/or correlation relationships were used in examining the relational nature of particular variables.

3.2 Participants

The questionnaire was applied, and 511 validated surveys were received, according to the demographic data in Tables 2, 3 and 4.

The participants in this study were selected through a non-probabilistic sampling process, using the convenience sampling method.

Considering these characteristics of the survey application, inferences from the sample results to the population using statistical tests, are not possible (Marôco, 2018).

The distribution of the male and female workforce shows a slight imbalance, 55% are female, and there is a strong feminisation of the less qualified professions (72% of women are production operators).

The respondents are mainly between the ages of 31–40 (41%) and 41–50 (32%), and there is also a group aged between 18–30, with 16.5% of the total sample. 59% of the respondents have more than 11 years of service length. Their service length ranges over 11 years, 75% of the total are female employees, while 43% of this group are male employees.

Regarding functions, women are more represented in the administrative and production areas. 90% of female employees belongs to the production departments,

Table 2 Characterisation of the sample (N = 511)

Companies	Valid answers	N
A	236	455
B	121	200
C	77	230
D	77	420
Total	511	1305

Table 3 Gender distribution (N = 511)

Relative frequencies (%)	
Gender	
Female	55
Male	45
Did not answer	0

Table 4 Demographic features (N = 511)

Relative frequencies (%)			
	Female	Male	Total
Age (years old)			
18–30	9	24	16.5
31–40	40	42	41
41–50	43	21	32
51–60	6	11	8.5
> 60	2	2	2
Service length in the company (years)			
< 1	4	3	3.5
1–3	7	14	10.5
4–6	4	22	13
7–10	10	18	14
11–15	18	17	17.5
16–20	23	9	16
> 20	34	17	25.5
Functions			
Administrative	15	14	14.5
Engineering	4	35	19.5
Management	5	13	9
Operators	75	20	47.5
Production technicians	1	18	9.5
Professional area			
Administrative and financial	4	2	3
Commercial and sales	2	13	7.5
Research and development	0	3	1.5
Maintenance	0	13	6.5
Production	90	65	77.5
Quality	4	4	4
Work contract			
No term contract	87	86	86.5
Fixed-term contract	13	14	13.5
Academic qualification			
< 7 years	23	1	12
7–9 years	27.7	8	17.8
10–12 years	40	36	38
1st cycle degree	8	45	26.5
Master's degree	1	10	5.5
Ph.D.	0.25	0	0.12

working as production operators, while male employees are distributed throughout all the functional areas of these organisations.

With regard to educational qualifications, it is the women (50.75%) who have the lowest level of educational qualifications (equal or less than 9th grade) in comparison to the men (9%). 38% of the employees have from the 10th to 12th grade qualifications, followed by 32.2% with a higher education degree. 29.8% of the sample has a 9th grade or lesser qualification, which reflects some of the responses to the specific characteristics of industrial production.

4 Result Presentation and Discussion

As mentioned above, the study aimed to verify the relationship between organisational trust, as defined in the reference literature and the variables gender and service length in the company.

4.1 Descriptive Statistics

To analyse the data, we started by calculating the level of reliability through Cronbach's alpha. The result was 0.875 (Company A = 0.878; Company B = 0.793; Company C = 0.907 and Company D = 0.890) which allowed us to continue the study, since the value was within the acceped limits (between 0.70 and 0.90) (Pestana & Gageiro, 2014), showing an adequate reliability of the reliability construct.

After that, to compare the mean of the quantitative dependent variable *"organisational trust"* according to the nominal independent variable gender, an ANOVA variance analysis was carried out (Table 5) and proved to be statistically significant.

The analysis was complemented by checking the ETA dependency ratio as shown in Table 6.

The value of 0.227 is closer to 0 than to 1, where 0 corresponds to the absence of a relationship and 1 to a perfect relationship.

Table 5 ANOVA[a] —organisational trust and gender

	Sum of squares	df	Mean square	Z	Sig.
Between groups	25,875	1	25,875	27,709	< 0.001[b]
In the groups	476,250	510	0.934		
Total	502,125	511			

[a]Dependent variable: organisational trust
[b]Preditor: (constant), gender

Table 6
ETA—organisational trust
and gender

	Eta	Eta squared
Organisational trust * gender	0.227	0.052

Therefore, in this case there is a relationship of association but not a very strong one.

The data in Table 7 show that confidence is on average higher among males. But could there be another variable involved in explaining the change in the degree of trust? We tested the analysis in age groups to try and understand if being older altered the degree of trust and in what way. The inexistence of statistical significance made it impossible to investigate this possible relationship.

After that, to compare the mean of the quantitative dependent variable "organisational trust" according to the ordinal variable seniority, an ANOVA variance analysis was performed (Table 8) and proved to be statistically significant.

After this analysis, the next step was to check whether or not service length is associated with organisational trust. In this case, since we are dealing with a quantitative variable (organisational trust) and another ordinal variable (service length), a Spearman's Ró correlation analysis was performed (Table 9).

The data show a moderate or low negative correlation with statistical significance, meaning that greater service length corresponds to less trust. Bearing in mind the questionnaire applied and the variable "organisational trust" we can only apparently see that increase in organisational knowledge, may make older people unwilling to become vulnerable to those who directly coordinate them in direct leadership.

Each organisation per se was not analysed in relation to the respondent's distribution by gender and service length, as but also by each organisation's results in terms of organisational trust for each group as far as gender and service length were considered.

Table 7 Gender—Descriptive analysis

Gender	Average	N	SD
Female	3.7827	283	0.98883
Male	4.2348	228	0.93779
Total	3.9849	511	0.99128

Table 8 ANOVA—organisational trust and service length in company

ANOVA

	Sum of squares	df	Mean square	Z	Sig.
Between groups	30,150	6	5.025	5.291	< 0.001
In the groups	500,500	527	0.950		
Total	530,650	533			

Table 9 Correlations result organisational trust and service length in the company

			Organisational trust	Service length in company
Spearman's Ró	Organisational trust	Correlation coefficient	*1*	*− 0.211*[**]
		Sig. (2 extremities)	.	*< 0.001*
		N	*548*	*534*
	Service length in company	Correlation coefficient	*− 0.211*[**]	*1*
		Sig. (2 extremities)	*< 0.001*	.
		N	*534*	*534*

[**], . The correlation is significant at the 0.01 level (2 extremities)

An attempt was made to disaggregate the analysis by organisation, but as there was no statistical significance for several of the companies in the study, it was decided to consider the aggregate data for the four organisations.

Even so, some information is provided in the following tables. Table 10 shows the distribution of respondents by gender in each of the companies.

Thus, Table 11 shows the degree of organisational trust according to the respondents' gender for the same companies and there is also evidence of more trust on men's behalf in all companies (A, B, C) except for organisation D where there were no differences based on gender.

Table 10 Distribution of respondents by gender in the companies

Company	Female	%	Male	%	Total
A	*195*	*82.6*	*41*	*17.4*	*236*
B	*20*	*16.5*	*101*	*83.5*	*121*
C	*28*	*36.4*	*49*	*63.6*	*77*
D	*40*	*51.9*	*37*	*48.1*	*77*
Total	*283*	*55.4*	*228*	*44.6*	*511*

Table 11 Organisational trust and gender by company

Company	Trust female	SD	Trust male	SD
A	*3.77*	*0.96*	*4.26*	*0.93*
B	*4.34*	*1.12*	*4.44*	*0.75*
C	*3.58*	*1.12*	*4.21*	*1.04*
D	*3.69*	*0.89*	*3.69*	*1.09*

Table 12 Service length in the different companies

		Company								Total	
		A	%	B	%	C	%	D	%	All	%
Service length in company (in years)	< 1	9	0.04	1	0.01	7	0.09	1	0.01	18	0.04
	1–3	15	0.06	13	0.11	17	0.22	9	0.12	54	0.11
	4–6	8	0.03	42	0.35	9	0.12	8	0.10	67	0.13
	7–10	19	0.08	22	0.18	8	0.10	21	0.27	70	0.14
	11–15	34	0.14	23	0.19	9	0.12	23	0.30	89	0.17
	16–20	67	0.28	2	0.02	3	0.04	11	0.14	83	0.16
	> 20	84	0.36	18	0.15	24	0.31	4	0.05	130	0.25
Total		236	1	121	1	77	1	77	1	511	1

Table 12 shows the distribution of employees by service length in each of the companies, with great heterogeneity in the organisations.

As shown in company A, 64% of the workers have at least 16 years of service length in the company. This company has the highest proportion of seniors, when compared to 17% in company B, 35% in C and 19% in D. If we analyse trust in each company in company A we see that the degree of trust decreases as service length increases.

As we analyse trust, we can see that in company A the degree of trust decreases as service length increases. It is also noteworthy that Company A, which has the most respondents, also holds a greater weight in the overall results. Although the pattern in Companies B and D, has some similarities, when we analyse the group of the oldest respondents the degree of trust increases (Table 13).

Table 13 Organisational trust and service length in each company

		Company				Total
		A	B	C	D	
Service length in each company (in years)	< 1	4.29	4.6	4.64	4.2	4.22
	1–3	4.16	4.46	4.38	3.45	4.22
	4–6	4.1	4.37	4.8	3.73	4.31
	7–10	4.2	4.48	4.18	3.91	4.19
	11–15	3.86	4.55	3.84	3.58	3.96
	16–20	3.69	3.96	4.35	3.55	3.74
	> 20	3.66	4.4	3.45	4.05	3.74

4.2 Limitations and Recommendations

To overcome some of the study limitations and as clues for future research, we suggest expanding the organisation's and respondents' number, as well as the variables under analysis, including some related to the characterisation of the supervisors/managers themselves. This may help clarify the knowledge about what promotes trust in organisations by the people who work in them.

These results therefore suggest that trust in these organisations is associated with gender, as male employees have a higher trust level in their supervisors and managers in general than female employees.

Women are mainly employed in production areas, as assembly line operators, and on average have lower educational qualifications than men. These differences in job qualifications and educational qualifications, may help understand the respondents' answers in future research.

5 Conclusion

A high degree of organisational trust enables organisations to achieve better performance from their employees, and also generates greater internal cohesion among teams.

Thus, it is important to understand what factors promote or contribute to improve the degree of organisational trust, which in the study case means the trust they have in their direct supervisors/managers. Trust relies on the premise that the one who trusts puts himself in a situation of vulnerability in relation to the direct supervisor/manager in this specific case.

Willingness to trust and to be in a vulnerable position naturally varies according, not only to the characteristics of both the person who trusts, and of the person who leads, but also to the way in which this leadership is exercised (e.g. leadership with more or less autocracy, more or less laissez-faire).

This research represents one of the few studies conducted in Portugal to empirically analyse the relationship between organisational trust and some individual characteristics, in particular gender and service length.

This study confirmed an association between degree of trust and gender, showing that the degree is higher when respondents are male.

Probably age or age group, the function one holds, gender, service length in the organisation, may have some a relationship within the organisational trust degree.

Probably due to the number of respondents, this study enabled us to verify statistical significance between organisational trust and some of these variables. We were able to identify the relationship of organisational trust with service length and gender. We confirmed that trust is higher for those with less service length and is higher in men than in women.

References

Barber, B. (1983). *The logic and limits of trust*. Rutgers University Press.

Barczak, G., Lassk, F., & Mulki, J. (2010). Antecedents of team creativity: An examination of team emotional intelligence, team trust and collaborative culture. *Creativity & Innovation Management, 19*(4), 332–345.

Bering, J. (2010). *The belief instinct*. W. W. Norton & Company.

Bhattacherjee, A. (2002). Individual trust in online firms: Scale development and initial test. *Journal of Management Systems, 19*(1), 211–241.

Bozic, B., Siebert, S., & Martin, G. (2019). A strategic action fields perspective on organizational trust repair. *European Management Journal, 37*, 58–66.

Colquitt, J. A., Scott, B., & LePine, J. (2007). Trust, trustworthiness, and trust propensity: A meta-analytic test of their unique relationships with risk taking and job performance. *Journal of Applied Psychology, 92*(4), 909–927.

Davis, J. H., Schoorman, F. D., Mayer, R. C., & Tan, H. H. (2000). The trusted general manager and business unit performance: Empirical evidence of a competitive advantage. *Strategic Management Journal, 21*, 563–576.

De Jong, B. A., Dirks, K. T., & Gillespie, N. (2016). Trust and team performance: A meta-analysis of main effects, moderators, and covariates. *Journal of Applied Psychology, 101*, 1134–1150.

Dirks, K. T. (2000). Trust in leadership and team performance: Evidence from NCAA basketball. *Journal of Applied Psychology, 85*, 1004–1012.

Dirks, K. T., & Ferrin, D. L. (2001). The role of trust in organizational settings. *Organization Science, 12*, 450–467.

Dirks, K. T., & Ferrin, D. L. (2002). Trust in leadership: Meta-analytic findings and implications for research and practice. *Journal of Applied Psychology, 87*, 611–628.

Dirks, K. T., & Skarlicki, D. P. (2009). The relationship between being perceived as trustworthy by coworkers and individual performance. *Journal of Management, 35*(1), 136–157.

Fukuyama, F. (1995). *Trust: The social virtues and creation of prosperity*. Free Press.

Garbarino, E., & Slonim, R. (2009). The robustness of trust and reciprocity across a heterogeneous US population. *Journal of Economic Behavior & Organization, 69*(3), 226–240.

Guinot, J., & Chiva, R. (2019). Vertical trust within organizations and performance: A systematic review. *Human Resource Development Review, 18*(2), 196–227.

Granovetter, M. S. (1985). Economic action and social structure: The problem of embeddedness. *American Journal of Sociology, 91*(3), 481–510.

Hardin, R. (2002). *Trust and trustworthiness*. The Russell Sage Foundation Series on Trust.

Hasche, N., Hoglund, L., & Martensson, M. (2022). Intra-organizational trust in public organizations-the study of interpersonal trust in both vertical and horizontal relationships from a bidirectional perspective. *Public Management Review, 23*(12), 1768–1788.

Hassan, M., & Semercioz, F. (2010). Trust in personal and impersonal forms its antecedents and consequences: A conceptual analysis within organizational context. *International Journal of Management and Information Systems, 14*(2), 67–83.

Hendriks, M., Burger, M., Rijsenbilt, A., Pleeging, E., & Commandeur, H. (2020). Virtuous leadership: A source of employee well-being and trust. *Management Research Review*.

Hosmer, L. (1995). Trust: The connecting link between organizational theory and philosophical ethics. *Academy of Management Review, 20*, 379–403.

INE. (2022). *System of integrated business accounts*. Instituto Nacional de Estatística.

Katou, A. A. (2013). Justice, trust and employee reactions: An empirical examination of the HRM system. *Management Research Review, 36*(7), 674–699.

Kee, H. W., & Knox, R. E. (1970). Conceptual and methodological considerations in the study of trust and suspicion. *The Journal of Conflict Resolution, 14*(3), 357–366.

Kramer, R. M. (2012). *Restoring trust in organizations and leaders: Enduring challenges and emerging answers*. Oxford University Press.

Krishnan, R., Martin, X., & Noorderhaven, N. G. (2006). When does trust matter to alliance performance? *The Academy of Management Journal, 49*(5), 849–917.
Lewicki, R. J., Mcallister, D. J., & Bies, R. J. (1998). Trust and distrust: New relationships and realities. *Academy of Management Review, 23*(3), 438–458.
Lewis, J. D., & Weigert, A. (1985). Trust as a social reality. *Social Forces, 63,* 967–985.
Maddux, W. W., & Brewer, M. B. (2005). Gender differences in the relational and collective bases for trust. *Group Processes & Intergroup Relations, 8*(2), 159–171.
Mahdizadeh, A., Hosseini, S. M., & Mehdizadeh, G. (2010). Identify the relationship between emotional intelligence and performance. In *2010 IEEE international conference on advanced management science (ICAMS 2010).*
Marôco, J. (2018). *Análise de equações estruturais: Fundamentos teóricos, software e aplicações* (7th ed.). ReportNumber.
Matzler, K., & Renzl, B. (2006). The relationship between interpersonal trust, employee satisfaction, and employee loyalty. *Total Quality Management and Business Excellence, 17,* 1261–1271.
Mayer, R. C., & Davis, J. H. (1999). The effect of the performance appraisal system on trust for management: A field quasi-experiment. *Journal of Applied Psychology, 84*(1), 123–136.
Mayer, R. C., Davis, J. H., & Schoorman, F. D. (1995). An integrative model of organizational trust. *Academy of Management Review, 20*(3), 709–734.
Mayer, R. C., & Gavin, M. B. (2005). Trust in management and performance: Who minds the shop while the employees watch the boss? *Academy of Management Journal, 48,* 874–888.
Mishra, A. K. (1996). Organizational responses to crisis: The centrality of trust. In R. M. Kramer & T. R. Tyler (Eds.), *Trust in organizations: Frontiers of theory and research* (pp. 261–287). Sage Publications.
Oliveira, A. F., Gomide Júnior, S., & Poli, B. V. S. (2020). Antecedentes de bem-estar no trabalho: Confiança e políticas de gestão de pessoas. *Revista de Administração Mackenzie, 21*(1), 1–26.
Pedro, L. (2015). *A GRH em contexto de crise: A centralidade da perceção de declínio organizacional* (Tese de doutoramento) ISCTE-IUL, Lisboa.
Pestana, M. H. & Gageiro, J. N. (2014). *Análise de dados para Ciências Sociais, A complementaridade do SPSS* (6th ed.). Edições Sílabo.
Podsakoff, P. M., MacKenzie, S. B., Moorman, R. H., & Fetter, R. (1990). Transformational leader behaviors and their effects on followers' trust in leader, satisfaction, and organizational citizenship behaviors. *The Leadership Quarterly, 1,* 107–142.
Robinson, S. L. (1996). Trust breach of the psychological contract. *Administrative Science Quarterly, 41,* 574–599.
Rodrigues, A., & Veloso, A. (2013). Organizational trust, risk and creativity. *Revista Brasileira de Gestão de Negócios, 15*(49), 545–561.
Rotter, J. B. (1967). New scale for the measurement of measurement of interpersonal trust. *Journal of Personality and Social Psychology, 35*(4), 651–665.
Rousseau, D., Sitkin, S., Burt, R., & Camerer, C. (1998). Not so different after all: A cross-discipline view of trust. *Academy of Management Review, 23*(3), 393–404.
Saparito, P. A., Chen, C. C., & Sapienza, H. J. (2004). The role of relational trust in bank-small firm relationships. *Academy of Management Journal, 47*(3), 400–410.
Schoorman, F. D. & Ballinger, G. A. (2006). *Leadership, trust and client service in veterinary hospitals* (Working paper). Purdue University.
Schoorman, F. D., Mayer, R. C., & Davis, J. H. (2007). An integrative model of organizational trust: Past, present, and future. *Academy of Management Review, 32*(2), 344–354.
Six, F., Nooteboom, B., & Hoogendoorn, A. (2010). Actions that build interpersonal trust: A relational signalling perspective. *Review of Social Economy, 68*(3), 285–315.
Uslaner, E. (2002). *The moral foundations of trust.* Nova Iorque.
Veloso, A. & Pinto, C. S. (2021). *Da psicologia à gestão de pessoas. Casos de intervenção em organizações.* Editora RH.
Webb, W. M., & Worchel, P. (1986). Trust and distrust. In S. Worchel & W. G. Austin (Eds.) *Psychology of intergroup relations* (pp. 213–228). Nelson-Hall.

Whitener, E., Brodt, S., Korsgaard, M. A., & Werner, J. (1998). Managers as initiators of trust: An exchange relationship framework for understanding managerial trustworthy behaviour. *Academy of Management Review, 23,* 513–530.

Williamson, O. (1993). Calculativeness, trust, and economic organization. *Journal of Law and Economics, 35,* 453–486.

Zaheer, A., McEvily, B., & Perrone, V. (1998). Does trust matter? Exploring the effects of interorganizational and interpersonal trust on performance. *Organization Science, 9,* 141–159.

Zak, P. J., & Knack, S. (2001). Trust and growth. *The Economic Journal, 111*(470), 295–321.

Zucker, L. G. (1986). Production of trust: Institutional sources of economic structure, 1840–1920. *Research in Organizational Behavior, 8,* 53–111.

Index

Printed in the United States
by Baker & Taylor Publisher Services